# HACKING WITH KALI LINUX

*A Step By Step Guide To Ethical Hacking, Hacking Tools, Protect Your Family And Business From Cyber Attacks Using The Basics Of Cybersecurity*

JEREMY DUNTON

© Copyright 2021 By Jeremy Dunton;

All Rights Reserved

This document is geared towards providing exact and reliable information in regards to the topic and issue covered. The publication is sold with the idea that the publisher is not required to render accounting, officially permitted, or otherwise, qualified services. If advice is necessary, legal or professional, a practiced individual in the profession should be ordered.

- From a Declaration of Principles which was accepted and ap-proved equally by a Committee of the American Bar Association and a Committee of Publishers and Associations.

In no way is, it legal to reproduce, duplicate, or transmit any part of this document in either electronic means or in printed format. Recording of this publication is strictly prohibited and any storage of this document is not allowed unless with written permission from the publisher. All rights reserved.

The information provided herein is stated to be truthful and con-sistent, in that any liability, in terms of inattention or otherwise, by any usage or abuse of any policies, processes, or directions contained within is the solitary and utter responsibility of the re-cipient reader. Under no circumstances will any legal responsibil-ity or blame be held against the publisher for any reparation, damages, or monetary loss due to the information herein, either directly or indirectly.

Respective authors own all copyrights not held by the publisher.

The information herein is offered for informational purposes sole-ly, and is universal as so. The presentation of the information is without contract or any type of guarantee assurance.

The trademarks that are used are without any consent and the publication of the trademark is without permission or backing by the trademark owner. All trademarks and brands within this book are for clarifying purposes only and are the owned by the owners themselves, not affiliated with this document.

## Table of Contents

- INTRODUCTION .......................................... 1
- WHAT IS ETHICAL HACKING? ...................... 5
- WHY DO ORGANIZATIONS NEED ETHICAL HACKERS 6
  - WHO IS AN ETHICAL HACKER? ................ 8
- WHAT SIGNIFICANCE DOES THE ETHICAL TRAINING OF HACKERS HAVE? ....................................... 10
- WHAT IS AN INFORMATION SECURITY SPECIALIST? 12
  - WHO SHOULD I PROTECT MY BUSINESS FROM? 13
- ETHICAL HACKING AND PENETRATION TESTS 101    14
  - HACKER ................................................... 27
- SKILLS EVERY HACKER NEEDS .................... 59
  - HACKING ................................................ 69
- HACKING RISKS FOR BUSINESSES ............. 73
  - WHO CAN BENEFIT FROM IT? ............... 77
- COMPUTER ETHICS IN THE WORKPLACE .. 80
  - ETHICAL HACKING ................................. 83
- TYPES OF HACKING AND HOW TO PREVENT THEM 87
  - DIFFERENT TYPES OF HACKING OVER THE YEARS 89
- DIFFERENT TYPES OF MORE ADVANCED HACKS    98
  - HACKING .............................................. 112
- HOW HACKERS WORK ............................. 117
- LEARN KNOWLEDGE ................................ 126

USEFUL BENEFITS OF YOUR ABILITIES. 126

ALWAYS TRY TO INTERRUPT THE CODE. 127

RECEIVING CERTIFICATION ................. 127

PROTECTING BUSINESSES FROM CYBERCRIME 138

KNOW YOUR DATA .............................. 148

TRAIN EMPLOYEES TO SPOT HARPOONS. 149

MAKE SURE THE SYSTEMS HAVE ADEQUATE FIREWALL AND ANTI-VIRUS TECHNOLOGY. .......... 149

HAVE TOOLS TO PREVENT DATA BREACHES, INCLUDING INTRUSION DETECTION. ..................... 150

INCLUDE DDOS SECURITY FEATURES. . 150

PROTECT YOUR BUSINESS WITH INSURANCE COVERAGE DESIGNED TO HANDLE CYBER RISKS... 151

WHY SHOULD SMALL BUSINESSES WORRY ABOUT CYBERSECURITY? ..................................... 153

PROTECTING YOUR FAMILY FROM CYBER ATTACKS 169

BENEFITS OF ETHICAL HACKING ............. 188

Development and quality assurance .. 190

Professional development.................. 192

Find Work ........................................... 193

Advantages of Ethical Hacking............ 194

Disadvantages of Ethical Hacking ....... 195

CONCLUSION.........................................197

# INTRODUCTION

Does the word hacking scare you? Generally, hacking has earned a negative reputation and has become associated with cyberattacks and breaches in cybersecurity. But this is not always true. If this is your first book on hacking, you will become more acquainted with the world of hacking as this book gives a simple overview of ethical hacking.

The term "ethical hacker" emerged in the late 1970s when the US government hired expert groups called "red teams" to hack their own computer system. Hackers are cyber-experts who lawfully or illegally hack. You enter the security system of a computer network to retrieve or recollect information.

Technology and the Internet have facilitated the creation and growth of network issues such as viruses, virus protection, hacking, and ethical hacking. Piracy is the modification of a hardware and software computer system. An illegal break in a computer system is a criminal offence. Recently, a wave of hacking has opened several courses on ethical hacking.

A White Hat Hacker is a moral hacker who performs intrusion tests and penetration tests. Ethical hacking involves lawfully hacking a computer system and entering its database. Its purpose is to hedge the shortcomings and drawbacks of a company's cybersecurity system. Legal hacking experts are typically certified ethical hackers who are hired to prevent potential threats to the computer security system or network. Ethical hacking courses are very popular and are considered by many as a serious profession. Ethical hacking courses have triggered huge reactions worldwide.

Moral hacking experts run multiple programs to secure corporate networks.

A moral hacker has legal permission to violate a company's software system or database. The company that approves an investigation of its security system must give its written consent to the School of Moral Piracy.

Moral hackers only deal with security issues in the enterprise and try to protect system vulnerabilities.

The hacker school performs a vulnerability analysis to address the vulnerabilities of the internal computer (PC) network. Also, they run software security programs as a necessary preventive action against illegal hacking.

Legal hacking experts identify security holes in a system that makes it easier for cybercriminals to get into it. They mainly perform these tests to ensure that the hardware and software programs are effective enough to prevent unauthorized access.

Moral experts carry out this test by simulating a cyberattack on the network to understand how powerful it is at interfering with the network.

The vulnerability test must be carried out regularly or annually. The company must record the results in full and review the references below.

Ethical piracy is an indispensable part of cybersecurity. The large and growing business activities of organizations as a result of globalization raise many security concerns that, if ignored, can lead to a huge loss of system integrity. therefore data theft. Every successful organization understands the importance of information security very well. Certified ethical pirates are therefore of great importance in this entire scenario.

# WHAT IS ETHICAL HACKING?

We all know and have heard about piracy, which is notorious in nature. Malicious hackers break systems and exploit them. Ethical hackers do the same, but for legitimate and legitimate constructive purposes, looking for vulnerabilities in the system and protecting it from attacks and potential threats. Ethical hackers provide information security and help businesses improve system security

# WHY DO ORGANIZATIONS NEED ETHICAL HACKERS

There are persistent cyber-attacks that have caused huge data losses and huge costs for their recovery. It is CE-Conseil, a leading IT organization for cybersecurity certification programs, which introduced the concept of ethical hacking for the first time since the September 11 terrorist attack. So far, ethical hacking has been accepted in the mixed computing world as it is more useful for securing systems and the network. Ethical hackers think just like malicious hackers and stop illegal activity by identifying threats and vulnerabilities.

The need for security experts or certified ethical pirates is greater than ever. Cyber threats keep ethical hackers on their toes to effectively protect the network and the system. No company can afford to lose confidential data. Although the concept of ethical piracy has been on the horizon lately, security concerns have already arisen and organizations have simply corrected it. Now, however, there is an urgent need to remedy the shortcomings, as the activities of organizations are becoming increasingly important.

Ethical hackers can be a savior for businesses in this whole business. Ethical hackers help organizations to do the following -

l Backup systems protect information from attacks by building a foolproof computer system that prevents illegal access

l Ethical hackers manage preventive measures to prevent hacking threats

l Create a safety awareness

l Regular network tests for periodic defense

Nowadays, doing business or personal activity online is not without deep-rooted challenges. Users have now begun to access other websites and accounts in order to falsify important information as well as to collect data. This form of stealth is called "hacking" and must be stopped! To prevent people from logging in to your account, you should check this regularly. This is the job of an ethical hacker. For this reason, the need for training in ethical piracy is increasingly being demanded worldwide.

## *WHO IS AN ETHICAL HACKER?*

In computer terminology, people with different intentions entering other people's accounts are identified as having hats of different colors! In this matter, the ethical hacker is a person wearing a white hat. The primary function of a person with computer hacking training is to enter or enter a system to verify the security and protection of the existing computer system. These white hackers are computer security experts and are trained to infiltrate a company's or individuals' systems that contain all the important and extremely sensitive information. It is important to note that such external systems seem very secure, but continue to face the dangers of a fraudulent approach.

In order to be trained in this activity, candidates are trained in ethical hacking. Companies use one of them or a group to verify the reliability of the system. When working as a team, they are called red teams or tiger teams. Through ethical training of hackers, they learn various methods to hack a system, and use these methods to sneak into gaps and to verify them. Many organizations now offer hackers ethical training programs.

# WHAT SIGNIFICANCE DOES THE ETHICAL TRAINING OF HACKERS HAVE?

Many companies are still not convinced that systems with ethical hacking need to be reviewed. They think nobody would hack their system. Ignorance could lead to loss the company millions of dollars. However, if the system is not checked by an expert and the information is hacked by bad people, very important and important information can get into the hands of the wrong person. Therefore, it is very essential to hire an expert with certified hacker training or to train employees internally. These individuals help identify and resolve problems in the system and protect the data from fraudulent use.

When these cheats are exposed to hackers, they also hack their employees' files and download viruses into a system that can shut down and damage the entire network. The consequences of such hacking can lead to the loss of important and secret information. It could be costly for the company for many of its customers if they no longer trust important information. These are only a little of the reasons why employing people trained in ethical hacking is of the greatest importance to the health of the company and its employees. After all, your IT health is in your hands!

With the rising use of computers and the Internet, users are facing more and more problems to watch out for. The ethical piracy training is to protect the site from fraudulent attempts to manipulate the main content and sensitive data.

Do you want your company to have a secure system? Protect them from suspicious persons who want to steal confidential documents by contacting a respected information security specialist. You can offer ethical hacking, intrusion testing, training, and consulting services to the PCI industry. Find out more about what it is and what it can do for your business.

## WHAT IS AN INFORMATION SECURITY SPECIALIST?

Another name for an information security specialist is a computer security specialist. This expert is qualified to protect the computer system from perils. These threats can be internal or external. In addition to private companies, the specialist provides services for public authorities and educational institutions.

The need for these professionals continues to grow. In fact, threats to computer systems and networks are rapidly evolving along with technological developments. For this reason, the skilled person must further improve his knowledge. It also needs to expand its arsenal of tools, applications and useful systems.

A basic security measure involves the control of passwords. An IT security specialist can ask members of the company to change their password frequently. This descreases the chances of unauthorized access to confidential programs, networks or databases.

## ***WHO SHOULD I PROTECT MY BUSINESS FROM?***

The most dangerous threat to a computer network is usually from the outside. The specialist sets up firewalls for hackers. It regularly installs programs that issue automatic alerts in an attempt to invade the system. You can even find high-tech programs to locate the hacker by identifying the intruder's Internet Protocol address.

The most popular IT services offered by computer scientists are ethical hacking and intrusion testing.

# ETHICAL HACKING AND PENETRATION TESTS 101

Qualified IT professionals usually run ethical hacking. They use their programming skills to understand the weaknesses of computer systems. As you find unethical hackers abusing vulnerabilities for personal reasons, the ethical hacker assesses and signals them, then proposes changes to strengthen the system. IT professionals provide system and information security with their ethical hacking services.

Most computer scientists regard hacking as mere hacking, as they always use the knowledge of computer systems to plant or invade them. Most business owners consider it ethical because of its purpose to increase the security of the systems.

Intrusion testing, on the other hand, is a type of security assessment performed on a computer system. A person tries to hack the system. The purpose of this service is to determine if a malicious individual can enter the system. Intrusion testing can detect programs or applications that hackers can access as soon as they enter the system. Many businesses and online businesses are offering intrusion testing. This is highly recommended because repairing damage to a computer system by a hostile attack can be expensive.

Most companies have to undergo intrusion tests. Adhering to the standard may seem difficult at first, but there are many companies that have the experience to meet the needs at all levels.

If you're looking for ethical hacking examples, read on!

It's funny, because the idea of launching a basically ethically offensive attack has certainly changed people's understanding of piracy. People tend to associate this immediately with negative actions and intentions because they only know the negative effects. In short, most people will believe that there can be little or no positive applications, but of course that is not true.

When hacking is well used, it is good!

As a means to improve the online defense of a person or company, we find this "malicious act" rather beneficial. The practice of breaking or bypassing an online system or network to expose weaknesses for future improvements is completely ethical (and you can earn a living as well).

Examples of ethical hacking are exploiting or opening a website to discover its weaknesses. Then report your results and have the person correct these vulnerabilities. If they are attacked in the future, then they are a bit safer. In fact, they prepare you for any threat from real attacks by eliminating areas that could potentially be exploited against them.

There are many examples of ethical hacking, including the one that took place in the early days of computer operation. At that time, the US Air Force used them to assess the security of an operating system. In this way, they were able to detect vulnerabilities such as vulnerable hardware, software and process security. They found that even with relatively little effort their safety could be circumvented and the intruder would get away with valuable information.

Through ethical hacking, they were able to prevent the occurrence of such an incident. The people responsible for this task treated the situation as if they were really the enemy and did everything in their power to penetrate the system. That way, they could see exactly how safe their system was. This may possibly be one of the greatest examples of ethical hacking as it has been sanctioned by those responsible for creating the online system. They have recognized the need for such measures because they know that there are many people who can do the same or do the same harm to their system.

Of all the ethical hacking examples, you may be clearly referring to known operating system practices that are in use today. Manufacturers of these operating systems make their own ethical hacking before presenting their products to the public. This is to avoid possible attacks by hackers. It is a kind of quality control during the system development phase to ensure that all vulnerabilities of its operating systems are covered, as it is marketed to the public. Ethical hacking is a very useful approach to defending your valuable online systems. By harnessing the capabilities and potential of White Hat hackers, you can combat and prevent the damage that real hackers do.

When people hear about hacking for the first time, they usually see this idea as something negative. In fact, hacking has always been to exploit websites or unprotected or weakly protected systems for the personal interests of all. Because of this, others (often companies) who want to improve the protection of their online systems are turning to professionals. These professional hackers (sometimes referred to as "white hats") use an ethical hacking method to strengthen their defense against actual hacking threats. By systematically "attacking" the system, they can quickly identify their vulnerabilities and then develop contingency plans to block, prevent, or eliminate real hacking attacks.

The ethical hacking methodology shows that not all hacks are bad. The fact of ethically hacking into a system to expose potential vulnerabilities that real pirates or "black hats" (due to less scientific intentions) can exploit can help prevent this from loss of income or bad reputation. In fact, many companies now seek the services of those who can perform this task because they understand that the only way to tackle skilled hackers is to hire another skilled hacker!

Those who have a thorough knowledge of computer systems can practice the execution of these services. Considering that ethical hacking methods involve intrusion into online systems, it is quite possible that many white hats today have their own experience as black hats!

As long as your actions have been authorized by the company that owns the system, the damage or disruption that you cause during the hacking process will be of full benefit to the company, provided these vulnerabilities are tracked and eliminated.

Ethical hackers are either hired professionals who have made a name for themselves as black hat hackers, or real company employees who have sufficient knowledge to perform the task.

They are not good or bad hackers, no white or black hats. In the end, it's about the benefits of society and the protection of the sensitive data it may contain. If you have a less desirable past (black hat), but have decided to work for the system rather than against it, you'll be well looked after by the service you can now deploy.

The method of ethical hacking is to produce results in protecting online systems from destructive attacks. You only care about protecting your assets and interests, and only when you think and act like a true kidnapper can you do so.

This is undoubtedly an effective way to protect yourself from online threats. If you're a company, do not hesitate to hire a white-hatred hacker who has the knowledge and skills to fight the threat of another hacker. On the other hand, if you were to hack yourself, would you have considered a career to work on the other side?

This process is performed by computer and network specialists known as ethical hackers or white hat. These individuals analyze and attack a company's security system to uncover vulnerabilities that can be exploited and exploited by hackers. It is important to understand that unlike hackers, ethical hackers are given permission by the competent authority to further test the security of their information system. Crackers cause damage and loss in an organization and affect the integrity, availability and confidentiality of an information system. How did the concept of ethical hacking come about and how does it work?

The field of ethical hacking has long existed in the computer world. Today, this topic has received much attention due to the availability and increasing use of computer resources and the Internet. This growth and expansion of the IT infrastructure has provided another opportunity for interaction, attracting large enterprise and government organizations. These organizations want to take full advantage of the technology to improve service quality for their customers. For example, companies want to use the Internet for e-commerce and advertising. In addition, governments want to use these resources to disseminate information to their citizens. Although they want to use this new technology-enhanced potential, there is a need for security. Companies fear that their computer system will be cracked and unauthorized persons will be able to access it. On the other hand, potential consumers and users of these services are concerned about the security of the information they are asked for. They fear that information such as credit card numbers, social security numbers, personal addresses and contacts may be accessible to intruders or outsiders using their data for purposes other than those intended. This disturbs their privacy, which is undesirable to many, if not all, people.

For the reasons mentioned above, organizations have been looking for a way to tackle and counteract this problem. They found that one of the best methods of limiting and controlling the threat to a security system by unauthorized personnel is to ask independent security experts to implement security measures a system. In this method, hackers use the same devices and techniques as intruders, but they do not damage the system and do not steal it. They evaluate the system and inform the owners of the vulnerabilities their system is exposed to. They also recommend what needs to be done on the system to make it safer.

As we can see above, ethical hacking goes hand in hand with increased security. Although this has contributed significantly to the increase in security issues, much remains to be done. It is impossible to get absolute security, but even if it is dangerous and undesirable, nothing to do for the safety of the computer.

Many people would be amazed to read about the advantages of ethical hacking. Such a concept does not exist for them, as piracy is automatically considered unethical or illegal. In fact, hacking usually involves removing the obstacles created for the protection and safety of human beings. Of course, talking about the benefits of such actions is quite alien to humans (at least in the beginning).

Initially, hacking was essentially about breaking the law and accessing information that some people normally do not have access to. But life is never just as black and white as we can perceive it first. That's why many people will be surprised that several large IT companies like IBM, Microsoft and Apple have a large team of dedicated hackers. Yes, you read that correctly!

However, they do not violate applicable law. No, these types of hackers are for good reasons. They are used as safety testers for all types of programs. Basically, every time a company offers a program, it communicates it to its hacker team, which then tests it ("hack") to determine the number of security flaws in the program.

You can see if the program can still be exploited, and then return it to the programmers with a list of weaknesses found. This is only one of the advantages of ethical hacking. The program can then be corrected or strengthened and returned to hackers to confirm if there are still problems.

The above is just one example of the benefits of computer piracy. Did you know that courses are being offered on this subject while demand for hackers has increased? With the world increasingly reliant on computers, the damage a hacker or hacking group can do has reached new heights. Big companies can not afford to ignore it.

Learning to be a hacker can lead to a promising career in which you work for one of the many big companies. As discussed earlier, there are numerous good reasons to conduct ethical hacker attacks "internally". They can help companies save potentially millions of dollars and minimize the risk of ruining their hard-earned reputation with their customers and their colleagues. This not only benefits companies, but also the people who buy their programs.

A team of efficient hackers can make sure that the program is as secure as possible, potentially hindering the work of any hacker and often forcing him to access simpler goals. This ensures that widespread programs are rarely manipulated, helping to protect the privacy and integrity of computers from people around the world.

# *HACKER*

The use of the term "hacker" to refer to a malicious computer expert is inappropriate. A hacker is someone with a high level of computer literacy, no matter how he uses it. There are three types of hackers: black, white and gray. Blacks are common, of course; Whites are good hackers; and gray, as the name implies, jumps between the two camps.

USA Today is interested in penetration testers ("pen"), also called ethical pirates. They are increasingly being hired by companies and security firms to see what works and what does not. The piece offers a good summary. The most interesting part of this article deals with the relative success of internal and external hacker attacks on the customer's office. The expert said he manages access to a company's internal systems almost 100% of the time, from 80% to 90%. Conversely, strong perimeter defense mechanisms reduce the pass rate by 20 to 30 percent when started on the other side of the firewall. Not least, it shows that the attention paid to the defense of the perimeter in recent years has borne fruit.

The potential benefits and significant issues surrounding intrusion testing are highlighted in a recent article in SC Magazine, which lists the recommendations of the National Institute for Standardization and Technology (NIST), which should include these procedures as a common federal instrument. The advantages are obvious: the penetration test can be used to detect and eliminate weaknesses from criminals or terrorists.

The drawback is that training people to do it like weapon training: there is no guarantee that knowledge will not be passed on to the source. Indeed, much of the article describes the control that must be exercised over these operations and the persons who perform them. NIST recommends involving outsiders to ensure that agency staff do not minimize the problems and reduce the risk of angered former employees organizing an attack. The suggestions will be finalized at the end of the month and published in March according to the story.

This is clearly an interesting and hot area. It seems that the quality of the pen tests is different and the chances are good that the field knows a lot of competition. This is a great overview of ethical hacking in Free Information Technology Tips. First, the author describes the contract called "Get out of Jail" card, as it frees the hacker of any criminal liability. This is necessary because much of what a hacker does is criminal. For an organization, it is very important to consult lawyers before hiring a responsible hacker. An obvious problem: If a company releases a hacker from lawsuits, does the liability remain if a customer sues, if the hacker made a mistake and data is lost?

The article describes three things he's trying to find: what information a hacker can get, what can happen to that information, and whether the organization automatically knows if a "real" hacker has planned a threat.

A regular look at ethical hacking in general and practitioner David Jacquet of Mainebiz show the rise of ethical piracy and explain why these people are in demand. The agonizing question concerns the boundaries between white, gray and black. The world of hacking is so exact, specialised and mysterious. How can companies be sure that the hacker they invite to attack their networks is really ethically correct? How do companies know that any weaknesses found have been reported to the customer?

Presumably, it's about reputation and trust. At the same time, it is a pretty difficult assumption.

Automated tools now determine the Internet. You can find some to expand your social networks, others to automatically respond to emails, and even robots to help your customers online. Of course, hacking has evolved too: today there are many automated OSINT tools that can help anyone carry out security investigations and information detection in a way that was not possible 20 years ago.

## 15 ETHICAL HACKING TOOLS YOU SHOULD NOT MISS

1. John the Ripper
2. Metasploit
3. Nmap
4. Wireshark
5. OpenVAS
6. IronWASP
7. Nikto
8. SQLMap
9. SQLNinja
10. Wapiti
11. Maltego
12. Air Cracking ng
13. Reaver
14. Ettercap
15. Screen

In recent decades, few security experts have conducted hacking tests and ethical intrusion tests. Now almost anyone can report security incidents. Using ethical hacking tools, you can analyze, search, and discover vulnerabilities and weaknesses in a company to increase the security of systems and applications (as described in the current Top CVE article for wild state released a few weeks ago).

Now, we will examine the best ethical hacking tools used by modern security researchers. We have compiled some of the well-known intrusion testing tools to help you in the early stages of a safety investigation. Here you will find classic tools that seem to have always existed, and new tools that may not be familiar to you.

1. John the Ripper

John the Ripper is one of the known password crackers of all time. It's also one of the best available security tools to test password security in your operating system or to remotely monitor it.

This password hijacker automatically detects the type of encryption used in almost all passwords and changes the password test algorithm accordingly. This makes it one of the several popular password hacking tools. Pass the smartest of all times.

This ethical hacking tool uses brute-force technology to decrypt passwords and algorithms such as:

Des, MD5, Blowfish

Kerberos AFS

Hash LM (Lan Manager), the system used in Windows NT / 2000 / XP / 2003

MD4, LDAP, MySQL (with third-party modules)

Another advantage is that JTR Open Source, cross-platform and fully available for Mac, Linux, Windows and Android.

2. Metasploit

Metasploit is an open source cybersecurity project that enables information professionals to use a variety of penetration testing tools to detect security vulnerabilities in remote software. It also acts as a development platform for operating modules.

One of the best-known results of this project is Ruby's Metasploit framework, which lets you easily develop, test, and run exploits. The framework includes a number of security tools that allow you to:

l Run systems for detecting hacking

l Perform security vulnerability scans

l Run remote attacks

l List networks and hosts

l Metasploit offers three different versions of its software:

l Pro: Ideal for intrusion testing and IT security teams.

Community: Used by Infoec small businesses and students.

Framework: The best for application developers and security researchers.

Supported platforms are:

Mac OS X

Linux

the window

3. Nmap

Nmap (Network Mapper) is a free, legal open source security tool that allows Infosec experts to manage and monitor the security of networks and operating systems for local and remote hosts.

Although it is one of the oldest existing security tools (introduced in 1997), it continues to be actively updated and receives new enhancements every year.

It is also known as one of the most efficient network imagery on the market, known for its speed and efficiency in delivering comprehensive results across all security investigations.

What can you do with Nmap?

l Audit of the safety device

l Identify open ports on remote hosts

l Network mapping and enumeration

l Find vulnerabilities in every network

l Start extensive DNS queries on domains and subdomains

Supported platforms are:

Mac OS X

Linux, OpenBSD and Solaris

Microsoft Windows

## 4. Wireshark

Wireshark is free open source software that lets you analyze network traffic in real time. With its detection technology, Wireshark is widely recognized for its ability to detect security issues in any network and efficiently solve common network problems.

Detecting the network enables you to catch and read the results in a human-readable format, making it easier to identify potential issues (such as low latency), threats, and vulnerabilities.

Main features:

Saves the scan for offline review

Package Browser

Powerful graphical interface

Rich VoIP analysis

Check and uncompress gzipped files

Reads other capture file formats, including: Sniffer Pro, tcpdump (libpcap), Microsoft Network Monitor, Cisco Secure IDS iplog, and more.

Supported network ports and devices: Ethernet, IEEE 802.11, PPP / HDLC, ATM, Bluetooth, USB, Token Ring, Frame Relay, FDDI.

Protocol decryption includes, but is not limited to, IPsec, ISAKMP, Kerberos, SNMPv3, SSL / TLS, WEP, and WPA / WPA2.

Exports the results in XML, PostScript, CSV, or plain text format

Wireshark holds up to 2000 different network protocols and is free on all major operating systems, including:

Linux

Microsoft windows

Mac OS X

FreeBSD, NetBSD, OpenBSD

## 5. OpenVAS

OpenVAS (also known as "Nessus") is an open source network scanner that detects remote vulnerabilities of all hosts. One of the most popular network weakness scanners, it is popular with administrators and professionals of DevOps and Infosec systems.

Main Features

Powerful web interface

+50,000 network susceptibility tests

Simultaneous analysis of multiple hosts

Ability to quit, pause, and resume analysis tasks

False positive management

Scheduled scans

Creation of graphics and statistics

Exports test results to plain text, XML, HTML, or LateX

Strong CLI available

Fully integrated with the Nagios monitoring software

The web interface makes it possible to work from any operating system. However, there is also a command-line interface suitable for Linux, Unix, and Windows operating systems.

The free version can be gotten from the OpenVAS website. However, a commercial corporate license is also available on the Greenbone Security (parent) Web site.

## 6. IronWASP

If you're thinking about ethical hacking, IronWASP is another great tool. It's free, open source and cross-platform and perfect for anyone who needs to monitor their web servers and public apps.

One of the most interesting aspects of IronWASP is that you do not have to be an expert to manage its main functions. Everything is based on a graphical interface, and complete analysis can be done in just a few clicks. If you start with ethical hacking tools, this is a great way to get started. Some of its key features include:

Powerful graphical interface

Ability to record the web analytic sequence

Exports the results in HTML and RTF format

Over 25 different vulnerabilities in the web

Wrong positive and negative management

Full support for Python and Ruby for the script engine

Can be extended with modules designed in C #, Ruby and Python

Hosted platforms: Windows, Linux with Wine and MacOS with CrossOver

## 7. Nikto

Nikto is another favourite known as part of Kali Linux Distribution. Other popular Linux distributions, such as Fedora, are already shipping with Nikto, which is available in their software repositories.

This security tool is useful for scanning Web servers and performing various test types on the specified remote host. The clean and simple command-line interface makes it easy to start vulnerability testing on your target, as shown in the screenshot below:

The main features of Nikto include:

Detects standard installation files on each operating system

Recognizes outdated software applications.

Run the XSS vulnerability tests

Starts brute-force attacks based on a warning

Exports the results to text, CSV or HTML files

Bypassing the Intrusion Detection System with LibWhisker

Integration with the Metasploit Framework

8. SQLMap

SQLmap is a cyber security tool developed in Python that allows security researchers to run SQL injection tests on remote hosts. With SQLMap, you can discover and test different types of SQL-based vulnerabilities to strengthen your applications and servers or report vulnerabilities to different companies.

The SQL injection techniques include:

UNION is based on a query

blind based on time

blind based on boolean

based on the error

stacked queries

outside the band

Main features:

Supports multiple database servers: Oracle, PostgreSQL, MySQL and MSSQL, MS Access, DB2 or Informix.

Automatic code entry

Password recognition

Password cracking based on a dictionary

Enumeration of users

Get password hashes

View user rights and databases

Increase database user rights

Information about the dump table

Run Remote SQL SELECTS

## 9. SQLNinja

SQLNinja is another SQL vulnerability scanner that ships with the Kali Linux distribution. This tool is intended for targeting and running web applications that use MS SQL Server as the primary database server. SQLNinja was written in Perl and is available in several Unix distributions on which the Perl interpreter is installed. It is supported on the following operating systems:

Linux

Mac OS X and iOS

FreeBSD

SQLninja can be executed in different modes, eg.

Test mode

Detailed mode

Remote fingerprint database mode

Brute force attack with a word list

Direct hull and inverted hull

Scanner for outgoing ports

Invert the ICMP shell

DNS Tunnel Shell

## 10. Wapiti

Wapiti is a free Python open source command line vulnerability scanner. Although not the most popular tool in the field, it can detect security vulnerabilities in many web applications.

Using wapiti, you can discover security holes, including:

XSS attacks

SQL injections

Injections of XPath

XXE injections

CRLF injections

Fake a server-side request

Some other features include:

Works in verbose mode

Possibility to stop and continue analyzes.

Highlights vulnerabilities in the terminal

Generate reports and export them to HTML, XML, JSON and TXT

Enable and disable multiple attack modules

Clears the parameters of some URLs

Exclude URLs during an attack

Ignore the SSL certificate validation

JavaScript URL Extractor

Set a time limit for large scans

Sets custom HTTP and user agent headers

11. Maltego

Maltego is the ideal tool for collecting information and recognizing data during the initial analysis of your goal.

In this case, relationships between people, names, phone numbers, e-mail addresses, companies, organizations, and social network profiles can be correlated and determined.

In addition to online resources such as whois data, DNS records, social networks, search engines, geolocation services, and online API services, this can also be used to investigate the correlation between Internet infrastructures. including:

Domain name

DNS server

Netblocks

IP addresses

Folder

Web pages


The main features include:

Graphical surface

Analyze up to 10,000 objects per diagram

Extensive correlation possibilities

Real-time data exchange

Graphical generator for correlated data

Exports graphics to GraphML

Generate entity lists

Can reproduce (copy and paste) information

This application is readily availablefor Windows, Linux and Mac OS. The only software requirement is the installation of Java 1.8 or higher.

## 12. Air Crack-ng

Air Crack-ng is a respected home security and enterprise security security suite. It provides full support for 802.11 WEP and WPA-PSK networks and captures network packets. He then analyzes them and uses them to interrupt the wifi access.

AirCrack-ng offers old-school security professionals an elegant terminal-based interface and some other interesting features.

Main features:

Complete documentation (wiki, manpages)

Active Community (forums and IRC channels)

Supports WLAN detection of Linux, Mac and Windows

Starts PTW, WEP, and fragmentation attacks

Supports WPA migration mode

Fast cracking speed

Support for multiple wireless cards

Integration with third-party tools

As a bonus, many Wifi Audit tools are included, including:

Air Force Base

Aircrack

airdecap-ng

airdecloak-ng

airdriver-ng

aireplay-ng

Air-ng

airodump-ng

airolib-ng

AirServ-ng

airtun-ng

easside-ng

paquetforge-ng

tkiptun-ng

Wesside-ng

airdecloak-ng

## 13. Reaver

Reaver is a great open source alternative to Aircrack-ng that lets you check the security of any Wi-Fi network with WPA / WPA2 passwords. It uses brute force Wi-Fi attack techniques such as Pixie's dust attacks to destroy WiFi-protected installations due to common security vulnerabilities and vulnerabilities in Wi-Fi.

Depending on the configuration of the wireless security on the router, it may take between 3 and 10 hours for the effective brute force to reach a cracking result.

Until recently, the original version of Reaver was hosted in Google Cloud. After the release of version 1.6, a forked community edition was released in Github. Some of the main features include:

Build up dependencies

essential construction

libpcap-dev

Runtime dependencies

Pixiewps (required for Pixiedust Attack)

It works well on most Linux distributions.

## 14. Ettercap

Ettercap is a network receiver and packet sniffer for LANs. It supports active and passive scanning as well as various protocols, including encrypted protocols such as SSH and HTTPS.

Other features include network and host analysis (such as the fingerprint of the operating system) and network manipulation over established connections. This makes it a proper tool for testing security attacks. Interceptor type.

Main Attributes

Active and passive protocol analysis

Filters depend on IP source and destination, Mac and ARP addresses

Inject data into existing connections

Protocols based on SSH and HTTPS encryption

Smells on long-distance traffic in the GRE tunnel

Expandable with plugins

Supported protocols are Telnet, FTP, Imap, SGB, MySQL, LDAP, NFS, SNMP, HTTP etc.

Determine the name and version of the operating system

Can terminate established LAN connections

DNS hijacking

## 15. Canvas

Canvas is a great alternative to Metasploit and offers more than 800 exploits to test remote networks.

Its main features include:

Remote networking

To target different types of systems

Target of selected geographic regions

Take screenshots of remote systems

Download passwords

Change the files in the system

Escalate permissions to gain administrator access

With this tool you can also use its platform to write new exploits or use its famous shellcode generator. It also includes an alternative to nmap called scanrand, which is especially useful for scanning ports and detecting hosts in medium to large networks.

Supported platforms are:

Linux

MacOSX (requires PyGTK)

Windows (requires Python and PyGTK)

Software vendors are leveraging automated ethical hacking tools and intrusion testing utilities to increase system security on a daily basis.

Automated tools are changing the way hacking evolves, making ethical penetration testing easier, faster, and more reliable than ever before. Intrusion testing and reporting activities now play an important role in identifying security vulnerabilities related to remote or local software. This allows business owners to quickly prevent security holes from spreading across the Internet.

Abilities enable you to achieve the desired goals within the time and resources available. As a hacker, you need to develop skills that will help you get the job done. These skills include learning programming, using the Internet, solving problems, and using existing security tools.

# SKILLS EVERY HACKER NEEDS

In this section, we show you the common programming languages and skills you need as a hacker.

What is a programming language?

Why should you learn to program?

Which programming languages should you learn?


What is a programming language?

A programming language is a language used to develop computer programs. The developed programs can range from operating systems. Data-driven applications through network solutions.

Why should you learn to program?

Hackers are the creators of tools and problem solvers. Learning programming helps you to implement problem solutions. It also differentiates you from script kiddies.

By writing programs as hackers, you can automate many tasks that usually take a long time to complete. Writing programs can also help you to identify and exploit programming errors in the applications you are targeting.

There's no need to constantly remake the wheel, and there are a number of easy-to-use open source programs. You can customize existing applications and add your methods as needed.

Which programming languages should I study?

The answer to the question will depend on your computer systems and platforms. Some programming languages are used to develop only specific platforms. For example, Visual Basic Classic (3, 4, 5, and 6.0) is used to write applications that run on the Windows operating system. It would be illogical if you would learn how to program in Visual Basic 6.0 if your goal is the abduction of Linux systems.

Programming languages for hackers

1 HTML

Language used to write web pages. * Multi-platform web hacking

Login forms and other methods for entering data on the Web use HTML forms to retrieve data. The ability to write and interpret HTML makes it easy to detect and exploit code weaknesses.

2 JavaScript

Client-side scripting language * Web hacking across multiple platforms

The JavaScript code is executed while the client is navigating. You can use it to read registered cookies and execute scripts between websites etc.

3 PHP

Server-side scripting language * Multi-platform web hacking

PHP is one of the most known and widely used web programming languages. It is used to receive HTML forms and perform other custom tasks. You can write a custom application in PHP that changes the settings of a web server and makes it vulnerable to attacks.

4 SQL

Language for communication with the database * Web hacking on multiple platforms

Use SQL Injection to bypass weak connection algorithms for web applications, delete data from the database, and so on.

5 Python

Advanced programming languages * Cross-platform creation of tools and scripts

They are useful when you need to develop scripts and automation tools. The insights gained can also be used to understand and adapt the tools already available.

6 C & C ++

High-level programming * multiplatform writing exploits, shell codes, etc.

They are useful if you need to write your own shell codes, exploits, rootkits or understand and develop existing ones.

7 Java
8 CSharp
9 Visual Basic
10 VBScript

OTHER TIPS

Java and CSharp are * multi-platforms. Visual Basic is particular to Windows Other Uses . The effectiveness of these languages depends on your scenario.

* Multiplatform means that programs developed with the respective language can be deployed on different operating systems such as Windows, Linux, MAC etc.

Other Skills A Hacker May Need

Besides programming skills, a good hacker must have the following skills:

- Know how to effectively use the Internet and search engines to gather information.

- Get a Linux-based operating system and master the basic commands that every Linux user should know.

- Practice makes perfect, a good hacker should work hard and make a positive contribution to the hacker community. It can help to develop open source programs, answer hacking forum questions, etc.

- Programming skills are crucial to becoming an effective hacker.

- Network skills are vital to becoming an effective hacker

- SQL skills are important to becoming an effective hacker.

Hacking tools are programs that make it easier to detect and exploit vulnerabilities in computer systems.

Over time, as we developed the technology and employed as a bee, we were able to develop and dramatically improve the state of the art. However, with the development of the technology, backlogs also remained, and this turned out to be a weak point and loops of the technology, which could lead to a hack situation where important information and data could be accessed on the Internet. Purpose theft, alteration or destruction etc.

Therefore these hackers have turned out to be the restless genius of information systems in which they can fight and harm the security activated on your device and put you at risk.

Recently, the need for information security has also increased as a person can help prevent such a situation and avoid apocalyptic moments. These people, who are the enemy of their unethical colleagues, have helped us to ensure the proactive security of information and to prevent us from having sleepless nights. They ensure the security of our data and information and protect them from catastrophes. We call them "ethical hackers". So what does it sound like to be an ethical hacker? Does it seduce you? If so, there are some points that you should consider and understand to become a professional hacker.

You need to recognize and understand the different types of hacking that can be divided into white hat, gray hat, and unethical hacking. You must rate all three to understand them well. Only thorough knowledge can help combat violations or unauthorized access to the information system. In addition, you can ensure proactive system security while identifying the vulnerability of information systems. Sound knowledge and powerful skills can help you become a good ethical hacker.

You must identify the basic requirements to be an ethical hacker. Whether you need a course or diploma or need a different license. Make sure you do your homework well and then practice.

After successfully assessing the baseline requirements, you need to guide the horses in your brain where you will need to know if you want to work with hardware or software. Believe me, this area is huge and you would like to take the risk of driving two boats at the same time. First, master one formula and then choose another one. So, choose it carefully.

Do not forget to use the UNIX operating system in addition to your traditional study or certificate program. It is the like the Holy Grail for hackers and the original operating system developed and designed by hackers only. Make sure you learn it well.

Once you have acquired your knowledge, it is time to know your strengths and weaknesses. Now try to apply your knowledge practically on your own system. Write down the full analysis and continue with another round. Go until you succeed and become highly qualified.

Last but not least; Identify the business aspect of your program. Discover job prospects or show off your own appearance. There are many lucrative jobs and positions in the market, both in the private and public sector, where you can make a living by facilitating services. Once you're done with it, the job is done.

Congratulations! You are very well on your way to becoming a professional hacker. Go, save the world.

## ***HACKING***

Computer hacking is one of the terms used in conversations to prove that the administration is facing computer security issues. Computer security breaches are reported on a daily basis and also occur in tightly controlled environments because users are not fully trained to identify them or critical systems are critical. Operations where you believe that all precautions have been taken will be hampered if a person steals important information in order to commit a crime.

It was not until November 2008 that it was reported that an inmate could access online employee files with personal information through a non-Internet computer. The programmers said access to the Internet had been prevented. However, "not for" does not mean much when integrated computer systems are used, as hackers find ways to bypass protected portals. It's like locking upthe front and back doors of a building, while the side windows remain unlocked.

- Understand the problem

In the case of the inmate, he accessed the employee files on the prison server with a thin client. Although the server was not programmed to access the Internet, the inmate skillfully entered the Internet by using the user and password information stolen from the employee files and discovering a portal. in software that is used by inmates for legal research purposes.

If an inmate can hack into a law enforcement system with sophisticated security systems designed to protect the public, there must be multi-level security that warns those monitoring the system that an attack attempt is being made. The goal is to detect and stop the error before information can be accessed. In other words, a well-designed security system has two characteristics:

* Security systems prevent intrusion

\* Trained employees with the necessary knowledge to recognize the signs of a hacking attempt and possible entry points for hacking.

You can use an Internet security service to assess your security and design an application that prevents intrusions. Employees who will use the system on a regular basis, however, need to know how hackers work and what methods are used to detect and use hackers exploiting vulnerable systems.

You have to know how to work

Basically, you teach your employees to become hackers to prevent hackers. Courses designed to familiarize employees with Internet security systems explain how hackers operate systems and how to detect them. They also learn how countermeasures work and return to the workplace to implement organizational measures to protect computer systems.

If the prison had set security levels that would provide notification that someone was trying to access employee files through software and prevented access, no violation would have occurred. It is crucial that your employees are able to identify potential vulnerabilities, detect hacking attempts, learn how to use business tools, and take countermeasures.

Often hackers provide confidential information all the way through, as employees do not recognize hacking. There is an old phrase that says, "You have to be known to know one." In the world of hacking, a highly skilled hacker needs to know a hacker. However, the benefit of this type of training is immeasurable, as the assets of the company are protected.

Hacking services helps protect your cybersecurity by providing world-class penetration testing and ethical hacking risk management services.

# **HACKING RISKS FOR BUSINESSES**

What risk do you see in hacking for an average company? If you own a small or medium-sized business, do you think you could ever be threatened?

It's tempting to think that the only companies that could really be hacked are banks, big wealthy organizations and maybe even governmental ones. But nothing could be more digressive from the truth.

While experienced and serious hackers attack the larger targets, there are still many intruders who attack the smaller ones, knowing that it will probably be easier to attack and damage them. If you do not take all the necessary steps to combat hacking, what you are doing is akin to going to work and leaving your front door with all your valuables on the dining table open.

Unfortunately, the same ignorance that seems to prevail over identity fraud sometimes also concerns hacking. We all would like to believe that this could never happen to us. The fact that you are sitting in a nice, warm and safe home office and making changes to your website on your own computer is not important to the hacker, who can cause you major problems, as well as being comfortable on the other side of the world.

There are several types of hackers, but they can all endanger your business. Some will try to steal personal information about their customers or conduct fraudulent transactions, while others simply want to cause as many problems as possible by attaching viruses to emails and sending them to as many people as possible. If someone who works for you opens this attachment, it can be the start of problems that keep your business in check until they are resolved.

In short, hacking is a business. And that means that your business can be seriously endangered if you do not admit that you are in danger. Do not ever make the mistake of believing that you are not at risk because you are not managing payments through your website, or even providing any kind of member area that a hacker can access. Everyone is at risk to a certain extent.

It is your job to make sure your business takes responsibility for its own safety, so to speak. You may have already implemented some basic security measures (firewall, anti-spyware, and anti-virus), and you must be aware of caution when opening attachments. But that does not make your business safe.

Consider instructing a specialist to assess your potential risk areas and how to combat them. Finding out how you and your company can work together to combat hackers is the first important step toward making improvements that hopefully hackers can not go beyond.

# CONSIDERING ETHICAL HACKING AS A CAREER

Are denial-of-service, intrusion detection, and hacking web servers your ideas for fun? If so, then it is the right choice to make a career as an information security professional or as an ethical hacker. An ethical hacker is also called a white-hat hacker who works legally to control the vulnerabilities of a computer or network. With the professional certification - Certified Ethical Hackers (CEH) you can be very much in demand on an organizational level.

What is a Certified Ethical Hacker (CEH)?

CEH is an ethical hacker that identifies network vulnerabilities and takes preventative measures to prevent data loss. A Certified Ethical Hacker Certificate (CEH) issued by the International Council of Electronic Commerce Consultants (EC Council) certifies the expertise of an information security specialist. This credential is vendor-neutral and accepted worldwide.

## *WHO CAN BENEFIT FROM IT?*

The course is very useful for people who work as site administrators, security professionals, auditors, and professionals with a network infrastructure.

How can you earn CEH?

The two means of obtaining this certification proposed by the European Council are:

Accredited Training Center (ATC): Anyone can join a five-day online or on-site training program at one of the centers. At the end of the process, the student or candidate can take the Prometric Prime exam on the Internet.

Self-learning: You can try the exam by searching the self-learning material provided and having at least two years of information security experience validated by the employer. If this is not the case, we can send you the details of the training with a request for consideration.

How do you sign up?

The steps to apply for this certification are:

- Fill in the form by receiving a confirmation from your employer.

- Include a copy of a government-approved passport, such as a passport.

- Send the scanned copy of the form and documents.

- Submit a non-refundable entry fee of $ 100.00.

The approval by the EC Council lasts usually two weeks.

exam details

Version 8 is the last certificate issued in late 2013. Candidates who pass the ATC exam mode will receive exam code 312-50. Candidates undergoing the self-study program can take the EC0-350 exam at a Prometric Authorized Test Center (APTC).

The exam consists of 125 multiple-choice questions. During the four-hour exam, candidates scan, test, hack and secure their systems. Candidates must receive 70% of the points to be able to express themselves with brio. Once the candidate is certified, the welcome kit will arrive after eight weeks.

# COMPUTER ETHICS IN THE WORKPLACE

Regardless of whether you are a highly skilled professional or a computer intern, you must adhere to computer ethics in the workplace. Computer ethics, like any ethical practice, is an established code of conduct. Otherwise, it can hurt others and cause you trouble at work and with the law. IT ethics is produced and regulated by a number of organizations, including the Computer Ethics Institute (1992) in America, and governments around the world. Although most have not been modified since the 1990s, they still apply today, even though the technology has evolved since then. Here are some key ethical issues related to today's computers.

Copy software

Over the past decade, the illegal copying of software, CDs and DVDs has become a sort of norm, but it is still illegal to do so. It is unethical to purchase software without payment, for which you have no permission to copy or use, as you do not pay yourself for not earning the deserved respect that the creator deserves for stealing your intellectual property.

Intellectual property

It is unethical to take the creation of another person and refer to them as yours, as this is the intellectual property of another person. When writing copies and in the ethics of intellectual property, this means that the product of the sale belongs only to them, and it's like theft if you credit it yourself.

Hacking

Whether you are hacking accounts or user IDs, it is still not ethically sound. At work, there are usually a number of computers or a multi-user credential network whose pirated content violates ethical and confidentiality rights. Computer hacking can cause many problems because it is essentially a computer that was deliberately designed to be private.

Data breach and destruction

Manipulating, manipulating, or intentionally destroying third party data without permission is unethical, as this would harm the user on purpose. If you also falsify data or steal personal information, you have violated not only the Code of Ethics but also the law.

Virus

Today, we spend most of our lives avoiding the dreaded virus that destroys or alters your computer. Damaging a computer, be it software, data or another computer, without permission is an illegal act. Under the Abuse of Computers Act in the UK, imprisonment or heavy fines are guaranteed.

Understanding and maintaining good computer ethics not only prevents you from interfering with the laws, but also helps maintain computer security. Understanding this Code of Ethics helps every trainee to understand that protecting computers and the data they contain is crucial in a workplace where private information is kept.

## *ETHICAL HACKING*

The formation of ethical hackers looks almost like an oxymoron. How can one be both ethical and pirate? You need to understand what an ethical hacker is, how he is trained and what he does to fully understand the genius of such a position.

The position is unique. The training teaches the same techniques that every hacker would try to infiltrate a computer system. The difference is that they do it to find weaknesses before they can actually be exploited. Finding vulnerabilities before they are open to the public can prevent actual intrusion into the system. Uncovering these vulnerabilities is just one way to test the security of a system.

Even though the hacking abilities are the same, it is the intent that makes the difference. Although these people could always try to invade the system to gain control over the internal functions of that system, they do so to find a way to protect that weakness. They identify the permeable areas in order to strengthen them. To stop a hacker, you have to think as such.

The training that such a person receives must be extensive. A thorough understanding of the way hackers invade systems is required so that the defenses taken are more than sufficient to stop any real hacker. If you forget a vulnerability in the system, you can be certain that there is an unethical type that exploits this vulnerability.

A variety of courses are offered to facilitate this training. A comprehensive network security course can prepare an interested person for work on the ground. This understanding of attacks and countermeasures is essential to the position. This includes knowing what to do in case of a system violation, investigating an attack attempt, and tracing cybercrime.

Ethical hackers are hired by a company to test the permeability of their network. Their efforts contribute to the security of information and systems in a world in which crimes associated with advanced technologies are becoming more prevalent. Finding network gaps is not an easy task, as attack and defense technologies are constantly changing and evolving at this level.

What was safe six months ago can now be hacked easily. Knowledge of the latest hacking techniques are fluent. It always changes. These skilled people perform risk analysis and help different areas of collaboration to ensure a high level of safety for the entire system. Those undergoing training even work on developing the new software, which will be implemented as soon as the vulnerabilities have been identified and the prevention measures taken.

The field of ethical hacker training will only expand as more and more business people accidentally or intentionally access publicly accessible computer systems. The security of corporate information, banking information and personal information is based on the ability to protect that information against external attacks.

This training leads a person to think like an external infiltrator, to stay one step ahead of the game, and the information they were hired to protect. Who knew that there is a good kind of hacker?

# TYPES OF HACKING AND HOW TO PREVENT THEM

Are you an aspiring hacker and want to learn more about different types of hacking?

This section covers all types of hacking and ways to prevent it.

Ian Murphy and Kevin Mitnick and Johan Helsinguis and Robert Morris. Do you remember these names for something?

Here's the catch, these are the geniuses behind some of the biggest hacks of all time.

Ian Murphy, also known as Captain Zip, is the first computer hacker to be convicted after hacking AT & T's computer network and changing internal clocks to measure billing rates.

From these small changes to massive exploits, the various types of hacking have disrupted the Internet, businesses, public networks, and more over time.

No one was bothered - Sony, Adobe, Target, Equifax, Yahoo, Marriot, eBay, JP Morgan Chase, LinkedIn and an endless list of big and small players were received.

Data Loss

Digital Marketing Course in Bangalore Source - Csoonline

The term "hacker" was coined in MIT's KI laboratories in the 1960s, and by the 1970s, the types of hacking attacks were already out of control.

Overall, these hacking techniques have become more sophisticated and targeted over the years. Computer hacking continued to be a major contributor to data breaches in 2018, accounting for 57.1 percent.

The threats continue to increase!

To understand what hacking is all about and how the various types of hacking have developed, let's look at it from the beginning to the day it became dangerous.

# ***DIFFERENT TYPES OF HACKING OVER THE YEARS***

Beginning of hacking (1960-1970)

All, because the term was invented in the 1960s, we regard it as the decade of the beginning of hacking. However, the fact is that the chronology of the computer security history of hackers points to some notable events as early as 1903, when the inventor Nevil Maskelyne disrupted the public use of Marconi's wireless telegraphy technology.

He sent insulting messages in Morse code and ignored the projector in the auditorium. Hacking took on a concrete form in the 1960s as programmers tried to hack systems for better results.

Those who did, turned against him and led to one of the worst types of hacking attacks of that time, when AT & T's wide-range change was broken off by whistle. high tone for free calls.The effects were so serious that the FBI had to intervene to crack the hacker network.

In 1979, Kevin Mitnick broke into his Ark computer system of Digital Equipment Corporation.

Middle Ages of Piracy (1980s)

With the introduction of PCs by IBM, every site was the subject of a hacker attack. From one million computer units in 1980 to 30 million in 1986, the PCs and networks they use have opened up new hacker opportunities. The gravity tip is acquired.

The Great Hacker War was announced by the Legion of Doom (LOD) and the Masters of Deception (MOD) hacker groups. Both groups worked as rivals by infiltrating private computers and computer networks, encrypting phone lines and monitoring calls illegally.

After the intervention of the Federal Criminal Police Office, arrests took place, the groups disbanded and the citizens learned about hacking.

<u>Piracy Intersects with Internet Walls (1990s)</u>

This is the time when hackers have been refined and adapted to the Web. The simple types of hacking programs have moved from BBS to hacking sites.

It began as a joke by computer programming addicts and has become a $ 2 trillion industry. Every computer and every computer network is a threat to attacks such as Denial of Service (DoS), Distributed Denial of Service (DDoS), MitM (Intermediate), Phishing and Spear phishing attacks. Password, SQL Injection Attacks, Cross-Site Scripting (XSS), etc.

Hacking attacks that have damaged the US Department of Justice, the CIA, the US Air Force, nuclear program systems, credit card websites, and all sorts of digital interaction sites have raised the question of what piracy is and what potential damage can be caused.

Possible damage due to hacking

This section discusses some of the types of hacking techniques you might encounter in 2019 as well as examples of these techniques. It explores the da,ages that may occur as a result of hacking.

Every personal information has become digital and our digital footprints have increased excessively. Unethical hackers attempt various types of hacking techniques to access data or computer systems that cause the damage.

Denial of Service (DoS \ DDoS)

GitHub, the open source coding development platform, was destroyed in 2018 with massive traffic of 1.35 terabits per second. This was the highest record she had ever received, and the platform was not ready to handle it.

A denial-of-service (DoS) is therefore designed to saturate the traffic of a website or network server far beyond what can cause a crash. Real users are faced with a denial of service. These are the most common types of hacking techniques.

As network owners continue to worry about troubleshooting traffic issues, sometimes a small malware program is introduced that takes several days to detect before the potential damage is done. Botnets are designed to automate the attack sending multiple requests, packets, or messages, which is overwhelming for the system.

A few years ago, distributed DDoS or DoS attacks were very simple. However, over the years DDoS toolkits have been able to launch both infrastructure and application attacks. It has been found that DDoS attacks from the application layer are very targeted.

### Key logger

A key logger or system monitor is software used to monitor and record keystrokes of a user on the keyboard or touchpad. The keystrokes are then retrieved to create the user's login ID and passwords without his knowledge.

The key logger can be software or hardware that steals personal information (PII). Anti-keylogging software can be used to detect the introduction of a key logger program or key logger hardware. The banking sector and other financial institutions are highly exposed to key logger attacks.

Public WiFi Fake WAP

An ordinary public WiFi user is the victim of a fake WAP attack in a café, an airport and in local shopping malls. The hacker creates a fake Wifi point, which is then connected to the free official WAP.

This is the simplest type of hacking attack that is possible, as a hacker must download simple software or know the "hot spot" feature of the mobile phone to create a fake WAP. The problem starts when hackers start to use a special data-gathering tool when the user connects to those spoofed WAPs.

With Fake WAP, an attacker attempts to steal your credentials or simulate a man-in-the-middle attack, such as Ettercap, that detects live connections or the Metasploit project from which the attacker completely controls the device. the WAP user.

Public WiFi Fake WAP

Fake WAP WiFi Source Public - IamWire

Attack of water points

Using a "hacker" hacking technique, the hacker tries to attack a certain group of internet users. This is achieved by identifying the website, which is mainly visited by the group.

The site is infected with viruses, worms, malware or malicious advertisements that allow an attacker to gain access to the user's computer system after the user has started the site. This is a kind of clickbait attack in which popular niche sites target users of the site. These are core attacks that infect a legitimate Web site, computer systems, or audience devices.

These types of attacks are committed in large companies, on bank websites, on government websites and in public places by setting false WAP and other methods.

The hacker constantly monitors the popular website with vulnerabilities. These sites contain malicious code that redirects the visitor to a malware web site waiting to infect the systems.

Waterhole attacks target organizations where employees, providers, or site users use less secure wireless networks to receive access. Although these attacks are not so common, they remain undetectable for a long time.

Attack of spring water points - Trend Micro

Often referred to as espionage attacks, a passive attack involves hacking that allows the hacker to listen to and easily monitor the transmitted information. In no case does a hacker attempt to change the user's data or system settings. It is a kind of espionage activity.

What is hacking like that? From a simpler point of view, this is the exact opposite of an active attack in which the hacker tries to enter the system by introducing malicious software and then changing the data. Although passive attack does not appear to be a major threat compared to an active attack, the damage can be significant if the hacker provides practical information such as your bank ID and password.

Many government agencies or high-end security organizations use passive attacks for non-malicious purposes. However, it has been found that actively attacking by gathering information using a passive attack is more targeted, making it difficult for the user to see that their information and data are being used to create a targeted attack.

# DIFFERENT TYPES OF MORE ADVANCED HACKS

Whenever a criminal tries to hack into a network, he tries to use the most common hacking techniques that have proven effective on many occasions.

He just tries not to reinvent the wheel and to optimize the existing techniques and perform more sophisticated and complex attacks. The cybersecurity attacks that make headlines today are usually one of the following:

Phishing

Phishing accounted for 90% of the data breaches and the number of phishing attacks increased by 65% in 2018 alone. The worst thing about phishing attacks is that they develop quickly. Every new month, 1.5 million new phishing sites are identified.

As a legitimate source, hackers use social engineering and various types of hacking techniques to trick victims into action that can range from clicking the link to downloading an attachment with a malicious link. Email is the most known channel for phishing attacks. Some types of highly targeted phishing attacks that are most commonly used are:

- harpoon Fishing

A well-planned attack in which information about the target group is collected and investigated in order to achieve a higher success rate. Disguised as a legitimate source, spear phishing email is not a random message that can be questioned. These are designed so that the victims trust them and cause a jerky reaction of the knee. Mostly these attacks occur during the holiday season, when email traffic is very high and the victims are lured into emergencies.

After the victim clicked on a malicious link, she used the login and password to log in. Then the hacker searched. Harpooning can be very dangerous

- Whaling

The whaling attacks are directed against the mailboxes of the leaders of a particular organization. CEOs and CFOs are most targeted. Since this management position still has access to confidential information, attacking this information is more beneficial for hackers. These include very targeted attacks. Individuals and organizations are all at risk.

- Viruses, trojans, worms

Software that enters the victim's system or network and continuously monitors or modifies the data and information transmitted. Viruses or Trojans are malicious and can track user data or encrypt the user's files by requesting the ransom to pass the decryption key.

Viruses, worms, Trojans and bots infect the computer system in other ways, but the effects depend on the kind of sophistication with which it is created.

- Cookie Theft

The theft of cookies is about imitating the cookie of a victim on the same network. By using these cookies, a hacker can gain access to websites that have already been visited by the victims, and malicious attacks can be carried out.

A cookie theft occurs when a hacker accesses unencrypted session data with which he acts as a real user. A user is exposed to cookies when browsing on trusted websites in public networks.

Hacking always leads to computer fraud where hackers try to access the user's personal information. The information can include email address, phone numbers, user IDs, passwords and even ID theft.

By accessing information on a personal level, hackers attempt to gain unauthorized access to bank accounts, social media accounts, and more to increase the damage. If it is not the direct route ID / direct password, the hacker tries to encrypt files, push malware and ransomware to extort the victim's money.

The development of the Internet was a game of good, evil and ugly, and "hacking" is the ugliest part. If you also want to become a cybersecurity expert, take the cybersecurity course and step up to a higher level of your career.

Most home and business offices now have a firewall that separates your internal computer network from the wild west of the Internet. The good news is that the firewalls have been configured more sophisticated and correct. This is a great job for backing up devices on your internal computer network. Modern firewalls now include detecting and preventing intruders, filtering emails, and blocking websites.

Most can tell who did what and when. Not only do they block criminals outside your network, they also prevent internal users from accessing inappropriate resources on the Internet. Employees may be prevented from visiting Web sites that waste your business or violate certain security requirements. The main working time is really not the time to update your Facebook page! Organizations do not want their healthcare and financial services staff to talk to a stranger via an instant messaging service.

The firewall is the electronic equivalent of the "gateway" to your computer network. There are a variety of potential perpetrators who paint their doors and windows, tirelessly searching for a way to get in. A properly configured, managed, and regularly updated firewall can effectively protect your computer network in the office and at home. Behind the firewall, desktops and desktop servers must have software-based firewalls based on local software that also provides antivirus protection. If anything goes beyond the firewall, then it's hoped that internal antivirus and desktop firewall solutions will provide an additional layer of security.

Firewalls are reasonable and adequate, but here are the bad news. Most of what you hear and read now is not caused by criminals crossing your firewall! The real damage is caused by those who are in your network! Malicious users and dishonest employees are always a boon. Unscrupulous employees give out credit card information or provide security information for money. The real danger, however, is that users will ignore the sophisticated security holes of today. The most honest employee may unknowingly become the source of a serious breach of security that results in the loss of your own personal information or your customers' personal and financial information.

Take the example of an average laptop user. How many times have you been to Starbucks and set yourself up? Have a nice day, open air, sun and broadband internet connection, cordless phone and the business runs as usual! If I tell you how stressless it is to organize an attack on Starbucks through a "middleman," you'll give up coffee for the rest of your life. You think you're on Starbucks Wi-Fi, but in reality, this kid on the back of the Starbucks, with the wireless access point connected to his USB port, gives the impression that he's your gateway to the Internet. It monitors every key on your laptop since you logged in. In fact, he now has your username, password, and almost everything else on your computer. If you return to the office and sign up now, you have just unlocked a bot on the corporate network, which will be available again this evening.

If laptops are not enough, everyone now runs with a smartphone! Did you know that your smartphone holds a list of all the Wi-Fi networks you've recently used? Do you remember when you waited at Starbucks for the cup of coffee and checked your emails? Now your phone sends a tag request anywhere that looks like "Starbucks WiFi, are you there?" in the hope that it will get an answer and automatically connect you to the internet. Do you remember the child we talked about? He decided to make your tag request with a "Yes, I'm here, hop!" To answer. Just another "MITM" attack and what it can do on your smartphone, especially the androids that make your laptop look like Fort Knocks!

When I sit in front of a door in a waiting room at the airport for fun and entertainment, I sometimes scan the WLAN to see how many phones, computers and iPads are online and connected. Do not say I would, but I think you could run a Netbios attack in less than five minutes? It's amazing how many people leave their network sharing options to their printer when they move. In the standard configuration, even more people are leaving their "network environment" settings! Exploration is always the same: Map the network to see which hosts are connected. Port analysis for known vulnerabilities; The toolbox of the exploit and the rest is getting boring for the ethical hacker. Well, credit card thieves, on the other side.

Your Internet browser is probably your worst enemy when it comes to protecting your privacy. Every website you visit, every email you send, and every link you follow are followed by hundreds of companies. Do not you believe me? If you're using Firefox, install an add-on called DoNotTrackme and see what's going on. Let's say you're an average user, and in less than 72 hours, you have a list of over 100 companies tracking every move on the Internet! These companies do not work for the NSA, but sell your "digital profile" to those who are willing to pay for the information. Where is your GPS? Which websites did you visit, which movies did you see, which products did you buy, which keywords did you select? All of this has been deliberately reported by you and your unsuspecting employees. Have you ever wondered if your competitors wanted to know what you saw online?

Voice over IP telephony systems provide a whole new set of security holes that must be exploited by unscrupulous criminals! Recently, we showed a law firm (as a paid intrusion detection and intrusion testing consultant and with the permission of its clients) how easy it is to secretly illuminate a law office based on a conference room and the entire conference via the Internet transmits a distant observer. ! In fact, capturing language packs for reading is the first joke kids learn in a hacking school!

VoIP, Bluetooth, WiFi, GPS, RFID, file and print sharing and even the "cloud" complement the list of vulnerabilities that can be easily exploited. What can you do? You must educate yourself and develop your own best practices for secure computing. You need to educate your staff and colleagues about the various vulnerabilities we face every day as we become more and more "wired" and mobile. Hire a competent security expert for computer networks to conduct "intrusion" tests on your organization's network and firewall. It would be easier to hire a specialist for "hacking" than to fix it after hacking! Remember, if we can touch your network, we will own it!

The formation of ethical hackers looks almost like an oxymoron. How can one be both ethical and pirate? You need to understand what an ethical hacker is, how he is trained and what he does to fully understand the genius of such a position.

The position is unique. The training teaches the same techniques that every hacker would try to infiltrate a computer system. The difference is that they do it to find weaknesses before they can actually be exploited. Finding vulnerabilities before they are open to the public can prevent actual intrusion into the system. Uncovering these vulnerabilities is just one way to test the security of a system.

Even though the hacking abilities are the same, it is the intent that makes the difference. Although these people could always try to invade the system to gain control over the internal functions of that system, they do so to find a way to protect that weakness. They identify the permeable areas in order to strengthen them. To stop a hacker, you have to think as such.

The training that such a person receives must be extensive. A thorough understanding of the way hackers invade systems is required so that the defenses taken are more than sufficient to stop any real hacker. If you forget a vulnerability in the system, you can be certain that there is an unethical type that exploits this vulnerability.

A variety of courses are offered to facilitate this training. A comprehensive network security course can prepare an interested person for work on the ground. This understanding of attacks and countermeasures is essential to the position. This includes knowing what to do in case of a system violation, investigating an attack attempt, and tracing cybercrime.

Ethical hackers are hired by a company to test the permeability of their network. Their efforts contribute to the security of information and systems in a world in which crimes associated with advanced technologies are becoming more prevalent. Finding network gaps is not an easy task, as attack and defense technologies are constantly changing and evolving at this level.

What was safe six months ago can now be solved easily. Knowledge of the latest hacking techniques are fluent. It always changes. These skilled people perform risk analysis and help different areas of collaboration to ensure a high level of safety for the entire system. Those undergoing training even work on developing the new software, which will be implemented as soon as the vulnerabilities have been identified and the prevention measures taken.

The field of ethical hacker training will only expand as more and more business people accidentally or intentionally access publicly accessible computer systems. The security of corporate information, banking information and personal information is based on the ability to protect that information against external attacks. This training leads a person to think like an external infiltrator, to stay one step ahead of the game, and the information they were hired to protect. Who knew that there is a good kind of hacker?

## *HACKING*

Social engineering / hacking can both be a means to compromise a person's or a company's private data, as well as a means to subsequently compromise.

One of the most common methods of attack is getting someone to use an infected USB key on their networked computer. I did it (without malware) just 2 days ago. I called my hotel reception and asked if I could send them a printable document because I forgot to bring a copy of my resume for a coroner's testimony to court. "No," said the clerk, "but do you have a jump order?" Then she put my flash drive into two different computers and turned the screen in my direction so I could ask her to open various files for printing.

Similarly, the Iranian nuclear research facility had infected an unauthorized USB reader with malicious software (Stuxnet) connected to a local computer. Although it is assumed that a duplicate Israeli agent has actually inserted one or more USB sticks into the site, it is not known if that person has left a few or if he or she has plugged in the hard disk itself. In any case, configuring a computer to fail to recognize a flash drive connected to the USB port is easy. That which is considered the very first successful cyberwar campaign can simply be the result, so that another person can grab another attractive device and use it alone with a remarkably unprotected system.

At the time of writing, Target Stores is in full-fledged reputation repair mode, as approximately 100 million credit cards have been compromised in its point-of-sale card readers. For several weeks, hackers rushed to a destination server every night, downloading thousands of credit card entries.

But how did the malware infect the server and POS devices? If we believe security researchers, the vulnerability probably comes from within. In the absence of an internal saboteur, a careless target employee, probably from the IT department, was misled by a link in an official e-mail - apparently from his bank, manager, or management. or visit a correspondent. attractive website - to reveal important permission information that has been passed on to the hacker. Or maybe someone has convinced a helpful employee to print a small document from a USB stick.

And now bankers are calling and investigating people to collect even more compromising information from unsuspecting victims as the second wave of social hacking.

Although there are sophisticated hackers (such as the author of the malicious software that eventually integrated them into Target's devices), this is the weakest link and therefore the least viable method for hackers is the reckless individual.

In 2008, 10 million carefree Americans had their credit card information stolen. In 2012 it was 15 million. In the last month of 2013 more than 100 million. Now some normative words in the wise:

DO NOT share your Social Security number, especially not by phone or email, and do not use it as an identifier. As a rule, it is sufficient to give it to your employer, your financial institution and the authorities.

With a few exceptions, when you're there, you DO NOT click on links embedded in e-mails -

especially links from people you do not know. It's probably a good idea not to click on the links of people you know. It's safer to manually enter URL or domain information into a browser. Do not share your passwords with third parties. Even the technical support would not need that. And do not be the helpful person who prints a person's CV by accidentally infecting the entire network.

There are courses that train ethical and unethical pirates in social engineering, and books like Hadnagy's "Social Engineering: The Art of Human Piracy" on the same topic. Many thousands of unethical characters take these lessons and are now looking for helpful souls like you. Take precautionary measures and do not let social engineers annoy you.

# HOW HACKERS WORK

A hacker or ethical hacker uses intrusion testing techniques to test a company's IT security and identify vulnerabilities. IT security personnel use the results of these intrusion tests to fix vulnerabilities, improve security, and reduce business risk factors.

Penetration testing is never an occasional task. This requires a lot of planning, including the explicit approval of management to conduct tests and then run tests that are as safe as possible. These tests often involve the same techniques that attackers use to actually break into a network.

<u>Background and training required</u>

White hat hacking involves a lot of problem solving as well as communication skills. A white-hat hacker also needs a balance between intelligence and common sense, strong technical and organizational skills, proper judgment, and the ability to remain under pressure.

At the same time, a white-hatred hacker must think like a black-hatred hacker with all his evil goals, abilities, and sneaky behavior. Some of the best white-hatred hackers are former black-hatted hackers who, for various reasons, have decided to leave a life of crime and use their skills in a positive (and legal) way.

There are no official training criteria for a white hacker - any organization can set their own requirements for this job - but a bachelor's or master's degree in information security, computer science or even mathematics provides a solid foundation.

For those who are not affiliated with the university, military training, especially in the field of intelligence, can help ensure that your resume is noted by human resource managers. Military service is also an asset to employers who want or need to hire people who already have a security clearance.

Relevant certifications

Many security certifications and security hackers can make it easier for a candidate to get started, even without much practical experience.

The CEH certification (Certified Ethical Hacker) by the European Council is a recommended starting point. The CEH is a vendor-neutral title, and CEH-certified professionals are in high demand. According to PayScale, the common salary of an ethical hacker is close to $ 80,000, and the upper range can reach more than $ 100,000. On the consultant side, the European Council states that CEH professionals can expect between $ 15,000 and $ 45,000 per contract or short-term mission.

Medium level CEH credentials focus on system piracy, enumeration, social engineering, SQL injection, Trojans, worms, viruses and other forms of attacks, including denial of service. Applicants must also have in-depth knowledge of cryptography, penetration testing, firewalls, honeypots, and more.

The CE Council recommends a five-day CEH training course for candidates without previous work experience. To pass the course, participants must have knowledge of Windows and Linux system administration, knowledge of TCP / IP, and basic knowledge of virtualization platforms. However, there are also opportunities for self-study to assist candidates in passing the required individual exam. Please note that the EU Council requires applicants to have at least two years of information security experience and to pay a registration fee of US $ 100.

Becoming a certified White Hat hacker also means staying on the legal side of hacking, never hacking illegally or unethically, and always protecting the intellectual property of others. As part of the certification procedure, applicants must agree with the EC Code of Conduct and never contact hackers or malicious activities.

Besides the CEH, the SANS GIAC program deserves a look. The organization has so far given more than 81,000 references. Candidates starting with the GIAC Cyber Defense Forums may be better able to complete active, respected, and comprehensive safety training. The GIAC Penetration Tester (GPEN) as well as the GIAC Exploit Search Tester and the Advanced Penetration Tester (GXPN) are notable arguments for aspiring hackers.

Another set of ethical hacking certifications comes from mile2. The organization's cybersecurity certification roadmap includes the Certified Vulnerability Evaluator (CVA), followed by the Certified Professional Ethical Hacker (CPEH), the Certified Penetration Testing Engineer (CPTE), and the Advanced Level Consultant for Certified Penetration Testing (CPTC). ). In addition, eligible US veterans can use their IG Bill benefits to earn cybersecure certifications and mile2 training.

Related Certifications in Forensic Science

It is always advisable to be tempted by computer forensics for those working in the field of information security. For those interested in the safety investigation component, continue with the EC-Conseil Certification List, then discuss the title of Computer Hacking Forensic Investigator (CHFI) certification.

CHFI focuses on the forensic investigation process and uses the appropriate tools and techniques to obtain evidence and computer data. As part of the CHFI certification training, candidates also learn how to recover deleted files, decode passwords, analyze network traffic, and use various forensic investigation tools to collect information.

Other notable forensic certificates include the forensically-certified GIAC Forensic Analyst (GCFA), the Forensic Forensic Technician, and the Computer Crime Researcher for the High Tech Crime Network.

## The physical side of the penetration tests

One last thing: all aspects of intrusion testing are not digital and are not always based on digital tracking methods or methods. Security experts generally refer to the security features of a site or facility, as well as the physical access controls associated with entering or personal use of facilities or equipment. the section "physical security". A full penetration test also attempts to jeopardize or circumvent physical security.

Trained intruder testers may try to get through an access gate, have someone close the door when attempting to bypass a badge reader or keyboard access control system, or use other forms of social engineering to circumvent physical security controls and barriers. Since it is first necessary to equip the equipment for attacks on their own security, physical security and the associated controls, procedures and security rules are just as important as similar measures on the digital side of the security barrier.

Most information security certifications, including CISSP, CISM, and Security +, cover the physical security in the general knowledge that candidates must learn and understand during preparation for testing. For people who are particularly interested in physical security, ASIS International's PSP (Physical Security Professional) card is probably the alpha and omega of physical security certifications. It is worthwhile to consult those who wish to understand the full range of penetration testing methods, approaches and techniques, especially in the field of physical security.

Candidates who want to work in InfoSec, as well as the relevant prior knowledge and one or two certifications, should not have any trouble finding an ethical hacking job immediately. Over time, you'll benefit from continuing education and certification to guide your career exactly where you want it.

From information security analysts to security managers, cybersecurity careers are in high demand. In particular, a profession is perceived by countless large companies.

Bring the hackers! There is a small difference between what a hacker and an "ethical" hacker do for their work. The hackers we know break in illegally and damage a system by exploiting its mistakes and weaknesses. Most hackers have a sneaky motive to make something out of it.

On the other hand, "ethical" hackers are trying to infiltrate systems to help the organization they work for, improve their security measures, fix bugs, and fix system weaknesses. So put on your "white hat" and get into the top 5 tips to become an "ethical" hacker.

## LEARN KNOWLEDGE

Discover the world of technology, from learning to coding in different programming languages, to understanding the operation of operating systems and networks. All of these common skills are needed and improved as an ethical hacker.

Gain experience through work. Years of work will naturally improve your skills and develop your professionalism in piracy. Remember, experience requires patience and effort to achieve your ultimate goal.

## *USEFUL BENEFITS OF YOUR ABILITIES.*

Being an ethical hacker does not just require manual skills. Hackers need to know different ways to damage firewalls and encrypt codes. Cybersecurity is an area where it is risky to work. You have to rely on more than just knowledge and experience. Hackers have to think "outside the system" to get into the system. In this way, improvements and security measures are being taken to protect networks and systems become more impenetrable.

## *ALWAYS TRY TO INTERRUPT THE CODE.*

Each time you pass an intrusion test, it is a step toward improving security and a step towards creating an impenetrable system. It will do you good for the general public.

## *RECEIVING CERTIFICATION*

You will not be hired by anyone who does not know your qualifications, even if you have the skills to handle a system. None of my aspiring hackers wrote me to demand the same. "What skills will you require to be a good hacker?"

The hacker is one of the most qualified disciplines of information technology, requiring a broad knowledge of technology and computing techniques. To really be a big hacker, you have to master many skills. Do not be discouraged if you do not have all the skills listed here, but use this list as the starting point for what you need to learn and master in the near future.

Here is my general list of the skills required to enter the elite pantheon. I've divided the skills into three categories to make it easier to move from one line to the next - basic, intermediate and soft skills - and links to articles over zero bytes that you can familiarize yourself with.

Basic knowledge

These are the foundational tips that every hacker should know before he tries to hack. If you understand everything in this section, you can move on to the intermediate stage.

1. BASIC COMPUTER SKILLS

It needs no saying that you must have basic computer skills to become a hacker. These capabilities go beyond the ability to create a Word document or surf the Web. What you need is to be able to access the command line on Windows, change the registry, and configure your network settings.

Many of these common skills can be acquired through a computer skills course such as A +.

## 2 NETWORK CAPABILITIES

You need to know the basics of network operation like the following.

DHCP

NAT

subnet

IPv4

IPv6

Public versus private IP

DNS

Routers and switches

VLAN

OSI model

MAC addressing

ARP

The better you understand how these technologies work, the more successful you are. Note that I have not written the following two guides, but they are very informative and cover some of the above basics of the network.

Foundations of hackers: a story of two standards

The worldwide guide to forwarding network packets

## 3 LINUX SKILLS

It is extremely important to develop Linux skills to become a hacker. Almost all of the tools we use as hackers are designed for Linux, and Linux gives us features we do not have on Windows. If you need to upgrade your Linux skills, or if you're new to Linux, you can read any of the Linux beginner series that are available online today.

## 4 WIRESHARK OR TCPDUMP

Wireshark is the most commonly used sniffer / protocol analyzer while tcpdump is a sniffer / command line protocol analyzer. Both can be extremely useful for analyzing traffic and TCP / IP attacks.

### An introduction to Wireshark and the OSI model
Wireshark filter for Wiretappers

## 5 VIRTUALIZATION

You need to master one of the virtualization packages, such as VirtualBox or VMWare Workstation. Normally, you need a safe environment to practice hacking before removing it from the real world. A virtual environment provides you with a secure environment for testing and tuning your hacks before you use them.

## 6 CONCEPTS AND SECURITY TECHNOLOGIES

A good hacker identifies and understands security concepts and technologies. The only way to override the obstacles created by security administrators is to know them. The hacker must include public key infrastructures (PKI), secure sockets (layers), intrusion detection systems (IDS) and firewalls. The inexperienced hacker can acquire many of these skills through a basic safety course, such as: B. Security +.

Reading and writing Snort rules to escape an IDS

## 7 wireless technologies

To hack wirelessly, you must first understand how it works. Things like encryption algorithms (WEP, WPA, WPA2), four-way negotiation and WPS. In addition, you must have knowledge of the connection and authentication protocol and the regulatory requirements of wireless technologies.

To know and learn more about the various types of encryption algorithms and examples of how to use wireless networking terms and technologies, read our collection of instructions on hacking Wi-Fi networks get operation of any piracy.

Get started with Wi-Fi terms and technologies

Instructions of the hacker hacker for hacking WLAN

## Advanced skills

It will be interesting and you will get an idea of your hacker skills. If you know all this information, you'll get more intuitive hacks where you hit every stroke, not another hacker.

## 8 SCRIPTING

Without scripting skills, the hacker must use the tools of other hackers. This limits your effectiveness. Every day, a new tool loses its effectiveness as security administrators take defensive measures.

To develop your own tools, you must master at least one of the scripting languages, including the BASH shell. This should include Perl, Python or Ruby.

## 9 BASIS OF KNOWLEDGE

If you want to be able to hack databases, you need to understand the databases and how they work. This includes the SQL language. I would also suggest or recommend using one of the leading database management systems such as SQL Server, Oracle or MySQL.

Prerequisites and technologies you need to know before you start

## 10 WEB APPLICATIONS

Web applications are possibly the most fertile ground for hackers in recent years. The better you understand how web applications and the databases they support work, the more successful you will be. In addition, you probably need to create your own website for phishing and other harmful purposes.

## 11 FORENSICS

You do not get caught to become a good hacker! You can not become a professional hacker who sits in a jail cell for 5 years. The more you learn about digital forensics, the better you can avoid detection and avoid it.

## 12 ADVANCED TCP / IP

The inexperienced hacker needs to understand the basics of TCP / IP, but to reach the intermediate level you need to understand the familiar details of the stacks and the TCP / IP protocol fields. This includes how each of the fields (flags, windows, df, tos, seq, ack, etc.) in the TCP and IP packets can be manipulated and used against the victim system to allow MitM attacks, among other things.

## 13 CRYPTOGRAPHIE

Although you do not have to be a cryptographer to be a good hacker, the better you understand your strengths and weaknesses, the greater your chances of overcoming any cryptographic algorithm. In addition, hackers can hide their activities using cryptography and avoid detection.

## 14 REVERSE ENGINEERING

Reverse Engineering allows you to open malicious software and rebuild it with additional features. Just like in software development, nobody creates a new application from scratch. Almost every new exploit or malware uses components of other existing malware.

In addition, reverse engineering allows the attacker to recover an existing exploit and change its signature to bypass IDS and AV detection. To change the signatures of the metasploit payload to avoid audiovisual recognition

Intangible skills

In addition to all these computer skills, the successful hacker must have immaterial skills. Some of these involve the following:

15 Reason CREATIVELY

There is ALWAYS a method to hack a system, and various ways to do it. A good hacker can think creatively about multiple approaches to the same hacking.

16 PROBLEM-SOLVING SKILLS

A hacker still faces seemingly unsolvable problems. This requires the hacker to be accustomed to thinking analytically and solving problems. This often requires the hacker to accurately diagnose the error and then split the problem into separate components. This is one of those skills that comes with many hours of practice. Problem solving is an essential skill of the hacker.

## 17 ENDURANCE

A hacker has to be persistent. If you fail at the beginning, try again. If that fails, suggest a new approach and try again. Only with stamina can you hack the safest systems.

Still undecided on hacking? A lot of information has been provided in this book to serve as a guide on what you need to learn and master in order to enter the world of ethical hacking and to give you information on protecting yourself form cyberattacks.

# PROTECTING BUSINESSES FROM CYBERCRIME

Cybercrime is becoming increasingly lucrative, outperforming other forms of crime. Unfortunately, there are only a few market entry barriers, ransomware is even available as a subscription service. The likelihood of being caught is also low, so the risk of grave consequences is not an effective deterrent. As businesses become more digital, increasingly vulnerable systems enable and manage critical services and all aspects of our daily lives.

The attack surface and the consequences of cybersecurity continue to increase alarmingly. The threat landscape poses an important problem that applies to all levels of the organization. How can you protect your business against cybersecurity threats?

Threats to businesses include attacks by experienced and advanced attackers. However, there are still numerous attempts with malware, ransomware, viruses, opportunistic hacking, social engineering and many other common threats. be a disaster for many organizations.

Regardless of the current state of your security program, there are ways to strengthen your defenses to counteract these attempts or mitigate potential harm.

1. Cyber security ratings

It is impossible to build up an adequate defense if you do not know where you are right now. Regular assessments of cybersecurity are a key part of any good security program, as they highlight the strengths you are building and the weaknesses that you can improve.

Getting started with an assessment will give you a clear idea of how you can protect your environment so that you can prioritize resources and avoid wasting time and money by wasting effort. If you've mastered a standard gap analysis, continue with the next cyber security assessment using a risk management approach. Set the right priorities and analyze the elements that you can put right.

2. Training of the staff

You are as strong as your least-informed employee. Attackers know how to find an entry point into their systems and network. It may be an unsuspecting employee with weak passwords or may be enthusiastic about phishing or social engineering. Make sure your team learns how cybercriminals can outsmart them, how to identify an email or a suspicious call, especially if they are legitimate friends or another service. Tell them how they can protect the organization from these attempts.

Remember to have strong policies that are aligned with cybersecurity best practices, and make sure your team knows them well. In addition, develop a team atmosphere as part of your awareness-raising initiatives. Encourage all team members to help each other and become an extension of your security team.

3. Keep the software up to date

Any software used by your company must be the latest version. Older applications are vulnerable to zero-day attacks and exploits that can steal information, penetrate networks, and cause serious harm.

Although antivirus software updates automatically, other programs may not have this feature. Regularly review all major software programs at least every two weeks. Consider implementing vulnerability management processes that are looking for missing patches and vulnerabilities. Then make sure you apply the required patches and updates.

4. Constant monitoring of threats

Attacks can happen anytime, anywhere. Implement 24/7 monitoring to stay alert and catch attacks before they cause damage. The Security Event and

Failure Monitoring (SEIM) software can alert you to suspicious user activity or data anomalies that may indicate an ongoing attack. The Operations Center's security services provide additional support in the form of security analysts who are trained in interpreting alerts and identifying people who report incidents. In addition, they know how to act quickly to stop an attack.

5. Emergency plan

With all the defense and protection features you implement, part of a global strategy is to assume that they will never be 100% effective. A well-designed and comprehensive incident response plan allows you to respond quickly and effectively if your organization suffers a successful attack.

A good emergency plan defines the appropriate escalation path. The best equipped team members will be notified immediately if problems occur. This will ensure that everyone understands the steps needed, who is responsible for what part of the response, and how to communicate with the organization's senior executives, external stakeholders, and the public when needed.

6. Less privileged access management

When evaluating the tools your team uses, make sure you've taken the time to define access permissions to team members, roles, levels, or roles. In short, ensure that only your administrators have access to all the features of a tool, system or network.

The rest of the mmebers of staff should only have access to functions, data and areas related to their work. When an attacker gains access to a person's credentials, the damage he or she can cause is limited to rights that are defined only for that person.

7. Physical security

When you focus on backing up your digital assets, you should not neglect the importance of securing your physical environment. In many cases, on-site attackers and physically entering buildings or data centers have access to credentials, trade secrets, infrastructure schemata, and other useful and actionable information.

Use proven security methods such as passport access, camera surveillance, and a visitor monitoring policy. Also, prepare your employees to protect their environment when out of the office by paying attention to remote work when traveling or visiting a local coffee shop. Maintaining calls and conversations, as well as private screens in these external environments, are just some of the tips you should consider when your employees are outside the secure area of the home office.

8. External Cyberintelligence

While it's invaluable to use cybersecurity ratings to constantly search for blind spots, features must be implemented that allow you to continuously monitor your systems, networks, and environment in the event of intrusion or interference. Suspicious activity is the same important to monitor networks, and external landscapes may affect you.

Cyber intelligence is an important part of an effective cybersecurity program. With Darkweb monitoring and other cyber intelligence methods, you can better understand what attacks are planned, which criminal networks may be targeting you, and how they plan to do so. You can also identify credentials or information that may already be in subterranean markets. With this knowledge, you can develop a precise defense strategy.

9. Assess the risks of third parties

Many important security breaches affecting major global brands have started with a vulnerability that was discovered by attackers at small vendors and third party vendors. Make sure that cybersecurity practices are part of your validation process if you want to work with a vendor. What are you doing to preserve and protect your networks, systems and data? What is your data destruction policy? Do you follow the rules that apply to you? Are you as alert as you are when it comes to controlling your employees? Implementing a security review process for your primary vendors, regularly updating your scoring, and integrating with incident coordination and threat monitoring as much as possible will help ensure that they maintain market share and do not update their business data. risky security.

10. Beware of theft of equipment

If your business uses mobile devices such as laptops, tablets, or other devices through a BYOD policy, you have a way to protect data remotely. In the best case you have to follow the devices. In the worst case, close the connection options when a device is stolen. At least make sure that encryption is used to protect the information that can be stored on the device. Although cybercrime mainly deals with digital information on the Internet, lost devices are always a possibility. An attacker can access a variety of information from a single stolen device. Prepare accordingly for the scenario.

Protecting your business against cyberthreats requires care and effective cyber security strategies. The combination of common sense and best practices, such as the tactics listed here, can help avoid attempts at cyber-attacks. Best of all, these solutions are cost-effective and you can usually spend less on cybersecurity because your activities are focused on the threats.

When a technology company was disturbed by a Distributed Denial of Service (DDoS) attack by a hacker who took control of one of his critical control panels, he was asked to pay for the resumption of control over his operations. The company decided not to comply with the blackmailers and instead tried to restore its account by changing the passwords. Unfortunately, hackers created backup logins in the panel and randomly deleted files as soon as they found the company's actions. The situation unfortunately brought the company out of business.

This cyber-blackmail blackmail scenario is becoming more common for all types and sizes of companies. One of the reasons for the increase in the number of incidents is that end-user software like Cryptolocker has downplayed the malware industry, making it more accessible to more criminals and less skilled hackers.

For extortion, cybercriminals may threaten to shut down computer systems or delete data, infect a company with a virus, publish confidential information or personally identifiable information about customers or employees, or initiate a security action denial-of-service attack or control take. Social media accounts.

Businesses can take the following eleven steps to protect themselves from cyber-blackmail:

## *KNOW YOUR DATA*

A business can not know exactly what the risk is before it understands the nature and amount of the data. Create file backups, backups, and bandwidth protection features. This helps a company to retain its information in the event of blackmail.

## *TRAIN EMPLOYEES TO SPOT HARPOONS.*

All employees must learn that it is important to protect the information they process on a regular basis to reduce business risk. Perform background checks for employees. Employee background checks can help determine if they have a criminal record. Limit the administrative capacity for systems and the social footprint. The fewer the employees who have access to confidential information, the better.

## *MAKE SURE THE SYSTEMS HAVE ADEQUATE FIREWALL AND ANTI-VIRUS TECHNOLOGY.*

After installing the appropriate software, evaluate the security settings for software, browsers, and email programs. Choose the system options that best meet your company's needs without increasing risk.

## *HAVE TOOLS TO PREVENT DATA BREACHES, INCLUDING INTRUSION DETECTION.*

Make sure people are really monitoring the detection tools. It is important not only to try to prevent a violation, but also to ensure that the company is informed as soon as possible when a violation occurs. Time is running out. Update security software patches on time. Regularly maintaining the security of your operating system is critical to its effectiveness over time.

## *INCLUDE DDOS SECURITY FEATURES.*

It is important to be able to avoid or ward off attacks that are likely to overwhelm or affect your systems. Set up a plan for handling data breaches. In the event of a violation, a clear protocol should be drawn up to identify the incident response team as well as their roles and responsibilities.

## ***PROTECT YOUR BUSINESS WITH INSURANCE COVERAGE DESIGNED TO HANDLE CYBER RISKS.***

Cyber insurance generally protects against data related to data breaches and blackmail. The right insurance program also gives qualified professionals access to the event from start to finish.

What would happen if a hacker started a cyber attack on your business? Would they be successful? Would you have easy access to your company's confidential information? Or would their attempt fail? Believe it or not, cybersecurity is not just about big business. This is something that small businesses need to look out for.

Consider the following small business security statistics:

43% of cyber attacks target small businesses.

Only 14% of small businesses believe that their ability to mitigate cyber risks, vulnerabilities and attacks is very effective.

Sixty percent of small businesses cease operations within six months of cyberattacks.

48% of the data breaches are caused by malicious acts. Human error or system failure counts for the rest.

If you own a small business, you can not ignore these statistics. You do not want your business to suffer because you have not taken the appropriate steps to protect it.

You've worked too hard to let your company be threatened by a hacker, right? In this article, you'll learn why it's important to focus on cybersecurity. You'll also learn how to preserve and protect your business.

# WHY SHOULD SMALL BUSINESSES WORRY ABOUT CYBERSECURITY?

I know what you are thinking. You think your business is so small that nobody ever wants to hack it. It's easy to imagine that a small business will never be confronted with cybersecurity issues. It makes sense, right? If you hear about a hacked company, it's usually a big brand like Target or Sony.

However, these are not the only goals.

This may be hard to believe, but hackers target small businesses as well. You just do not hear about it because the media will not talk about piracy with small businesses. There are several reasons why a hacker attacks a small business and they include:

- Small businesses do not take cyber security seriously: Let's be honest. Most small business owners do not take cyber security seriously. They consider themselves too small to attract the attention of a hacker.

However, this is one of the major reasons why a small business can be hacked. Hackers know that most small business owners do not invest in cybersecurity. Why? Because small business owners think they have nothing worth stealing from. This makes it a simple goal. There is a good chance that hackers are looking for something: the payment information of the customers.

- They have information that hackers want: Your company may not be as big as Target or Starbucks ... but it does not matter. You pay for your products and services, right? It means that you have something that hackers want. You have the payment information of your customers. You have information from your employees.

The Council of Better Business Bureaus revealed that 7.4% of small business owners had been cheated. As a business owner, you have information about customers and employees. This information is as valuable to hackers as gold. If your system is not secure, these hackers may have access to payment information and social security numbers. You have to make sure that this information is protected.

How to Protect Your Small Business from a Cyber Attack

So I have shown you that a small business does not necessarily mean that you can not get hacked. But if you are intelligent - and I know - you're probably wondering how to protect your company's information. That's what the next section of this article is about.

Complete a cyber insurance

Insurance is not just for your car, house or medical bills. You can also take out insurance for your business. In fact, every business should take out some kind of business insurance.

But there is also a cybersecurity insurance. If you are a small business, you need it. Of course we all hope that there will be no security breaches. But hope is not enough. You have to make sure your business is covered.

Cyber liability insurance is designed to prevent your business from a variety of cybersecurity threats. In the wake of a security breach and corporate liability, you may be required to pay a large sum to the court. This can paralyze most small businesses. If you have a cyber liability insurance, you do not have to worry about it. If you purchase the right type of insurance, your legal fees will be covered.

Develop a password strategy

Many cybersecurity attacks occur because the passwords your employees use are too simple. If your team is not trained, it may use passwords that are too easy to hack. It happens all the time.

For this reason, you need to create and set up an effective password strategy. You may not be able to stop any attack, but you can certainly slow down a stubborn hacker. If your system is not easy to hack, it can discourage the attacker. You will move on to another small business owner who is not as smart as you!

Luckily it's easy enough. You need to make sure that your team members have passwords that include uppercase and lowercase letters, numbers, and symbols. Yes, I know it can be painful, but the security of your business is worth it. In addition, you must ask your employees to reset their passwords at least once a month.

Use Virtual Datarooms (VDRs)

Virtual Datarooms are an excellent way to secure your business information. They facilitate the exchange of sensitive data by your employees. A virtual data room is an online library where your company can store data. They are typically used in financial transactions. It will prove very hard for an attacker to access information stored in a VDR.

A company can store many types of information in a VDR:

Financial information

Legal documents

Tax documents

Information about intellectual property

VDRs are a great way to ensure the security of your sensitive information.

- Talk to an expert

Yes, most businesses avoid doing this. But you should not. The payment of a computer security adviser may seem a bit expensive. But it is a great investment.

If your home is leaking and water is accumulating in your bathroom, would you try to repair it yourself? Probably not. You would probably call a worker, right? Why? Because if you are like a majority of us, you do not know what plumbing is about. The same principle applies to computer security.

If you're worried about cybersecurity, which you should be, consider talking to a computer security expert. An IT security consultant can review your business and set the best course of action to protect against cyberattacks.

An IT security consultant can point out areas where your business is vulnerable to cyber attacks. You can make recommendations that help protect your business. When it comes to the area of cybersecurity, you should never be too careful. If this is part of your budget, hire an expert. They will be glad to have done it.

- Beware of internal threats

This may come as a big reveal, but most cyber security issues arise when someone is in the business. It's not an idea most business owners want to think about, but it's absolutely true.

Here is a hard truth: 55% of all cyber attacks come from the company. 31.5% are made by malicious employees. 23.5% comes from insiders of the company who accidentally leave an attackable company. The protection of your business lies in the organization. It is easy to assume that a cyberattack emanates from an external force. But that's not correct. You need to focus on both the people in your organization and those outside of your organization.

Be sure to monitor your authorization requirements. Be careful when deciding which employees should have access to confidential information. This will help prevent "internal piracy".

Do not feel guilty about watching the activities of your employees. As the proprietor of your business, it's your job to protect yourself and your team. I have understood. You do not want micromanaging. The utmost is to find a balance between safety and playing big brother. It's different for every business, but if you work there, you'll find that balance.

- To sum up everything

If you own a small business, you need to take your cyber security seriously. Do not assume that your business is not an end just because you're not a big company.

You owe it to your team and your customers to ensure the safety of your company. Preventing cyber attacks should be amongst your top priorities. If you take the right action, you do not have to worry about putting your business at risk.

Businesses of all sizes are now the target of cybercriminals. According to Verizon 2018 DBIR, 58% of victims of data breaches are small businesses. In addition, it is shocking that, according to the National Cyber Security Alliance, 60% of small businesses are closed within six months of an attack. What makes these small businesses vulnerable to cyber attacks? It is probably the lack of resources due to the limited budget and the false conviction that only large organizations are attacked by pirates.

However, businesses of all sizes must defend themselves against cyber attackers. I've developed some useful preventative measures to protect your business from cyberattacks.

- Train your employees

Your employees are your greatest tool, but also the biggest security risk. Your main action should therefore be to make your employees aware of safety. This helps to minimize cases of accidental or intentional data loss. An important point to keep in mind is that training your staff is not a one-time task. This should be carried out regularly to ensure that your employees are kept up to date on the latest cyber threats. This helps them cautiously handle security vulnerabilities and threats.

- Manage your passwords

Your passwords are the access to your company's confidential information. It's important to follow some basic rules for creating and managing passwords for your business.

Always change the default passwords to very unique passwords

Avoid creating and using the same password for different accounts

Make sure your passwords are securely stored. Use a password manager. Never write your passwords on paper that others have access to.

Follow the instructions to create a strong password. Use a concoction of uppercase and lowercase letters, numbers, symbols, and so on.

- Keep your technology in shape

The operating system and enterprise system applications must be up-to-date, as this ensures the installation of the latest security patches. In addition, a firewall and virus protection must be installed on each system. Make sure they are both active, up-to-date, and installed with the correct settings. Microsoft operating systems have a standard firewall. You have to activate it. However, it is highly recommended to invest in reliable and advanced PC antivirus software. After all, buying an antivirus program costs much less than being a victim of a cyberattack.

- Keep backups to limit losses

With the increasing number of ransomware attacks, the importance of data protection became clear. It's best to keep a copy of your data instead of taking the risk of paying a ransom to the hackers. A business can resume normal business after an attack if a backup is available. Perform regular backups of your company data as this facilitates recovery from a current point in time. Backups must always be kept on a separate system.

- Have periodic tests for your coding

The code and hosting of your website is an important aspect of the security of your business. Have your website completely tested by your internal information security team for security errors or rent one. Incorrect or outdated code can help hackers access your website and cause damage. Also, make sure that your website hosting is from a reliable hosting company. Remember to get a security certificate for your website because it protects harmful content on the web at a reasonable price.

- Obtain an Endpoint Security Solution (EPS)
Complete endpoint protection gives you peace of mind that you have control over device, application, and web security. An EPS file provides protection for all endpoints in your enterprise that attackers use to begin malware. EPS preventive solutions for small and medium businesses require a small investment. As a business, however, you must consider the value of the security that such a solution offers.

## WHAT TO DO IN CASE OF INFRINGEMENTS?

If you have already been cyber-attacked, you must do the following:

Do a check of everything

First of all, you need to know the type of data lost or stolen during the data breach. Is it a generic form of data such as addresses, names, etc.? Is it more sensitive data such as e-mail addresses, credit / debit card numbers? Or are the lost data extremely sensitive, such as: For example, passwords, online banking credentials, credit / debit card security codes, etc.?

- Make the changes immediately. Take action

Depending on the type of data loss, you need to take action. For example, if passwords are lost, you must immediately change all system credentials, the Web, and other credentials. Ensure that you do not repeat the passwords for two accounts and that they are completely different from the lost accounts. To add an additional level of protection, you can use two-factor authentication. Even if a cyber-thief has the correct password, he can only enter if the security code is generated by two-factor authentication. If the lost information contains online banking credentials, it is important to notify the bank or the card issuing authority as they may suspend your card to prevent unauthorized transactions. In addition, they could release a new card to ensure your safety.

- Inform about the violation

Your customers should be aware of the violation of their business. So once you know that your business data has been compromised, share it with your customers. Tell them what kind of data has been compromised, what the company is doing to limit the losses, and what security measures can be taken.

- Perform a gap check

It is essential to conduct a post-breach audit to identify the root cause (eg inadequate security at the terminals) and identify the cause of the violation. You can hire an InfoSec reviewer to help you understand the depth and depth of the attack, and provide recommendations for improving your data processing practices.

- Last but not least - do not panic

Violations are certainly stressful events and panic is a natural reaction to such incidents. But I recommend that you do not panic. This will take a lot of time, which is better spent on planning actions.

Since it is predicted that cybercrime will cost the world about $6 trillion by 2021, it is necessary to consider the security practices outlined above for your company's security.

Cyber security for the whole family: Five surprising ways to get hacked by children and adults

# PROTECTING YOUR FAMILY FROM CYBER ATTACKS

Cyber security is a risk for children and adults. Learn how you can protect your family from identity theft and cyber-attacks at home and abroad.

When you are online, you are visible around the world. With more than 10 billion devices connected to the Internet, there are plenty of hackers. In 2017 alone, 179 million records were uncovered and today one in 15 people are victims of identity theft, including children.

Do you have online accounts? Wi-Fi compatible devices? A social security number? In that case, it's a good idea to pay attention to your back as hackers could choose you and your family. But do not panic for now. Knowing what to search for can drastically reduce the risk of cyberattacks.

You've probably already heard about the basics of cybersecurity. These include constantly changing passwords and utilizing a password manager, running periodic software updates, and refraining from opening a password. Attachments of people you do not know. While cybersecurity experts learn to prevent attacks, criminals are only inventing new ones. So let us identify some of the new ways in which hackers break in and how you can preserve your family from cyber-attacks. Two thirds of the cyber attacks cases are not solved.

1. More and more sophisticated phishing scams

Most people believe that they are not attacked by a cyberattack, but two-thirds of the attacks are not targeted. That's why cyber security is so important to families. Email phishing scams is a great example. In fact, 92% of cybersecurity issues start with emails, so the whole family needs to take precautions, especially when sharing computers.

End the time of obvious phishing. Do not expect princes from exotic countries to send you an e-mail asking for money. Now hackers mimic us by claiming to be the ones we trust - banks, mobile companies, governments, and even friends and family. When opening emails, be aware of the following warning signs:

2. Request private information

If you're asked for bank account numbers, passwords, or other sensitive information by email, it's probably a phishing scam. Scammers know that e-mail comes from authoritative sources such as the IRS, but organizations like the IRS do not send e-mails or request further information in the event of a problem. will contact you by US post or come directly to your door.). If you're not sure who the email or caller is, look up his number on Google (do not use the ones listed in the suspicious email) and call him to find out.

Note: If you receive harassing phone calls or emails from someone claiming to be in the IRS or any other government agency, report it.

3. Exceptional emails

If a friend's email account is compromised, a hacker can contact you by email and see if he or she can access your accounts. If you receive an e-mail from someone you know you do not expect, just call the person - a hacker might be able to impersonate someone in an e-mail, but it's unlikely that he pretends to be to be a voice, not to mention the phone's access to his victim.

Unusual language, typos and account changes

Be careful with subtle hints, such as emails from jane.doe@gmail.com, if your contact is listed as jane-doe@gmail.com. Business emails are unlikely to contain typos or strange words. A sentence that you do not know well from someone you know well may be a red flag. Please read it carefully.

You may be asked to click on an attachment or link

You might have heard that, but it's worth repeating. If you do not know what you are opening, do not open it until you talk to the person who sent it. If an e-mail comes from your bank asking you to connect using a link provided, protect it - avoid the link in the e-mail and simply log in to your online bank account in the usual way. Why? Because you never know what that stubbornness might be. You could very well download malware that can access the entire hard drive. If you work on a shared computer, you are putting all users at risk.

What to say to your family: Trust, but check and call before you click. A call takes two minutes. Eliminating the damage caused by a cyberattack can take years, even if you recover completely.

2. steal the identity of children

We hear a lot about cyber security among seniors. They always seem to be victims of phishing scams by email and phone. But when it comes to identity theft, the likelihood of children being affected is much higher. Why are children a preferred destination? Think about it. Children do not have to check their creditworthiness before the age of 18. Criminals can take credit cards in their name and live big. You will accumulate unrecognized debts until your child is denied a student loan because it has about $ 1 million in your pocket.

The red flags for these can be easily recognized. If you receive a letter from the IRS stating that your 8-year-old has not paid any taxes or has collected calls for items that you have not purchased, follow these instructions to learn the steps to take if your child's identity is stolen becomes. Courtesy of the FTC. However, since cybercriminals often wait several years before using the information they receive, you may not notice red flags. Remember to check your children's annual free credit reports to make sure everything looks good. In addition, you can specify a credit stop (also known as a safety stop) in the credit reports of all your children. Freezing credit is one of the easiest ways to protect children from identity theft. This free tool prevents anyone from accessing a credit report. Since most new accounts will not be approved without this report, identity thieves will find it much harder to use your information.

What to say to your family: Tell your children that you will freeze (or at least monitor) their report to prevent cybersecurity issues from affecting them in the future. When you review your credit report, involve them in the process to familiarize yourself with financial management.

3. Access confidential information using unsecured devices connected to the Internet

Does anyone care what you put in your Wi-Fi-connected slow cooker? Of course not, but that does not mean it will not get hacked. Intelligent home appliances are incredible, but the benefits come with risks.

The Internet of Things is a cybersecurity risk, because if a hacker can access a single device, they can access your entire network. This application-controlled baby video monitor can be a serious drawback to your digital armor. Many of the common cybersecurity policies apply here. Use passwords that you regularly change, try multi-factor authentication, perform security updates, and avoid using public WLAN. However, whenever possible, avoid connecting to the Internet with devices that can not be used safely. Ask yourself if this device really needs to connect to the Internet, and in that case, try placing it on a different network than your main computer.

<u>What you have to say to your family:</u> Sorry, kids, we do not have a Wi-Fi compliant refrigerator, but we have a password manager.

4. Enjoy social sharing on social media

It is quite normal to want to talk about your vacation. This is exciting, you will create many memories, take lots of pictures and it is not something you do every day. But if you reveal on Facebook that you are leaving the city for two weeks, you may also have a sign saying "Please, steal my house" hanging on your door.

It's okay to have public social media sites like yours. However, avoid sharing personal information on these pages. Keep personal pages private and review your friends list from time to time. Are you really in contact with 1,000 people? Remove people you do not know, personalize your security settings to limit the number of people seeing your information, or stop sharing personal information in that profile. This may seem like patanoid steps to take but they are not any less valid. One of the popular ways hackers use to break into security systems is simply looking through the social media accounts of their targets.

Please note that with every security update, Facebook will reset your default settings to the lowest security setting. Check the settings regularly and use secure passwords. Many social platforms even offer multifactor authentication. This means that you will need to use your password and one or more additional verification forms to access an account. For example, this could be a one-way code sent to your phone or a fingerprint - anything a hacker may not be able to use.

What to tell your family: Set private pages, update security settings regularly, delete unknown people, and avoid sharing your location information. When traveling, remember to wait for your return home before publishing the photos.

5. Start cyber security against travelers

Staying away from social networks as you leave the city can help protect your home while you're on the move. What can you do to protect your information when the plane lands? Just as there are hackers at home, there are also hackers abroad - and travelers are simple and unsuspecting targets.

Suppose you stay in a hotel and want to get some emails for work. You open your laptop and see several Wi-Fi options: Hotel-1, Hotel-guest and Hotel_publicwifi. Pick the wrong one, and you might be in a fraudulent Wi-Fi hotspot exposing your information to an unpleasant character who is responsible for the trap.

Criminals set up dummy access points and wait for you to log in so they can see everything you do on the Internet, for example, by logging into your online banking account. The more visitors there are, the greater the risk:

Disney, Paris, New York, the Walk of Fame, and the Las Vegas Strip are teeming with deceptive Wi-Fi hotspots.

Protect your family against cyber attacks by sticking to personal hotspots or introducing a clean device (a device that does not contain any of your information) instead of a computer with years of financial data.

Also, take precautions when using cards. Flying over counters may seem like an old story, but it's on the rise again. Criminals can use keyboard overlays or place card collectors on the actual DAB card reader. However, they usually place cameras over the keyboard so they can see your

PIN. Bank-linked ATMs are usually safe as the bank can register regularly. If possible, stick to them and cover your hand when entering a PIN. It does not hurt to shake the card reader as well. If it's loose, you might be dealing with a card skimmer.

Older children and adults who do not know the technology are likely to make these slight mistakes while traveling. Make sure that the whole family understands the risks.

What to say to your family: Be careful when traveling by using a device or a credit / debit card. Store them at bank-connected ATMs and shake hands when entering a PIN code. Do not connect to the public Wi-Fi network and instead use Wi-Fi hotspots or personal information (or, even better, store devices and enjoy your vacation).

And now?

Most compromises come in minutes, but only 3% are discovered so quickly. The detection of attacks on two-thirds takes months and is almost always discovered by third parties. It may seem like a big effort, but it is much easier to prevent an attack than to recognize it and reverse the damage as soon as it happens. That's why it's so important for you to share online security with your entire family.

Here are some of other mistakes you may be making that leave you vulnerable to cyber attacks:

1. Open emails from unknown persons

E-mail is the preferred form of business communication. According to the Radicati Group, the average person receives 235 e-mails every day. With so many e-mails, it goes without saying that some are frauds. Opening an unknown email or attachment in an email can trigger a virus that opens a backdoor for cybercriminals in your company's digital home.

Solutions:

Advise employees not to open e-mails from people they do not know.

Advise employees never to open attachments or unknown links.

2. Have weak credentials and passwords

Mashable reported that 81% of adults use the same password for everything. Repeated passwords that use personal information such as a nickname or a street are a problem. Cybercriminals have programs that extract public profiles for combinations of potential passwords and logon opportunities until one of them appears. They also use dictionary attacks that automatically try different words until they match.

Solutions:

Employees must use unique passwords

Add numbers and symbols to a password for added security. For example, replace "London" with "L0nD0N".

Create rules that require employees to create unique complex passwords of at least 12 characters. and change them if there is reason to believe that they have been compromised.

Reduce the problem by using password management software to automatically generate secure individual passwords for multiple applications, Web sites, and devices.

3. Leave passwords on sticky notes

Have you ever gone to the office and found a reminder with passwords on a screen? It happens more often than you think. Although you want a degree of confidence in your organization, it is too risky to leave passwords visible.

Solutions:

If employees need to enter their passwords, request that the printouts be kept in locked drawers.

4. Have access to everything

In some cases, companies do not subdivide the data. In other words, everyone, from the trainee to the board member, can access the same company files. By giving everyone the same access to data, you increase the number of people who can flee, lose information or manipulate information.

Solutions:

Set hierarchical access levels and grant permission only to those who need them at each level.

Limit the amount of people who can change system configurations.

Do not give your employees administrator rights on their devices unless they need such a configuration. Also, employees with administrative rights should use them only when needed, not routinely.

Apply a double signature before payments over a certain amount can be processed to combat tax fraud.

5. Lack of effective employee training

Research shows that the majority of businesses offer cybersecurity training. However, only 25% of executives believe that training is effective.

Solutions:

· Annual cybersecurity awareness. Topics could be:

Reasons and importance of cybersecurity training

Phishing and online scams

Lock the computer

Password Management

How to manage mobile devices

Relevant examples of situations

6. Do not update the antivirus software

Your company needs to provide antivirus software as a safeguard, but it is not up to the staff to update it. In some organizations, people are encouraged to perform updates, and they can decide whether to perform updates or not. Employees are likely to reject updates when they're in the middle of a project because many updates force them to shut down programs or restart computers.

Antivirus updates are important, have to be processed quickly and should not be left to the employees.

Solutions:

Automatically configure all system updates after hours.

No employee, regardless of title, should subscribe to this company policy.

## 7. Use of unsecured mobile devices

Do your employees have a cell phone, tablet or laptop? If so, did you set up a protocol to back up these devices? Many companies are careless about mobile devices, but they are an easy target for cybercriminals.

Solutions:

Each device must be password protected.

If a device is misplaced or stolen, contact a contact point to report it and specify the steps to remotely disable it.

Use endpoint security solutions to remotely manage mobile devices.

Do not conduct confidential transactions over an unreliable public Wi-Fi network.

Employees are human and digital accidents can happen. However, taking steps to protect equipment and train employees can help prevent cyber threats.

Of course, cybersecurity in your business goes beyond employee training. Protecting the digital footprint of a business and managing threats requires the help of a reputable cybersecurity company.

# BENEFITS OF ETHICAL HACKING

To learn more about ethical product piracy, hacker and penetration tester attitudes, tools, and tools need to be explored to learn how to identify, sort, and fix security holes in software and computer networks. The study of ethical product piracy can be useful to employees in a variety of roles, including network lawyers, risk management, software developers, quality assurance, management and legal reviewers. In addition, continuous ethical product piracy training and certification may be beneficial to those seeking a new role or demonstrating skills and values for their organization.

Understand the mentality of hackers

The most practical benefit of learning ethical hacking is its potential to inform and improve the way a business network is defended. The biggest threat to enterprise network security is a hacker:

knowing how hackers do it can help network lawyers identify, sort, prioritize, and determine how to best address potential threats.

Advocates of the network are extremely disadvantaged by hackers. A hacker simply has to identify and exploit a vulnerability to gain a foothold in a network, while a defender must theoretically identify and resolve any potential internal security and network perimeter vulnerabilities.

In practice, it is difficult to fully eliminate all the risks of a network, and a defender must be able to evaluate the likelihood of exploitation and the expected impact of each potential threat and allocate limited resources to minimize the likelihood of attack. successful. To be successful, a defender must be able to think like a hacker. Ethical hacking training can help a network advocate develop this state of mind.

## ***Development and quality assurance***

The roles of Ethical Hacker and Quality Assurance Tester overlap greatly. In both cases, the tester must ensure that the software functions properly under normal and extreme conditions. With today's fast development cycles, security testing is often overlooked, making the software vulnerable. A qualified ethical hacker could be an important resource for a development team that can perform security testing quickly, efficiently, and comprehensively, using best practices across the industry, rather than developing internal methods that put too much emphasis on some things and others to neglect.

Beyond learning security best practices, learning ethical hacking is also helpful from the tools point of view. Many cyber defense lawyers, quality assurance testers and hackers have developed tools to accelerate the identification and resolution of common security vulnerabilities. When a developer becomes familiar with and masters these tools, he learns which coding mistakes to avoid and how he can effectively test for security holes in the code.

Compliance with legal regulations
Recent regulations have taken a much stronger position in corporate responsibility for data breaches. The new provisions of the General Data Protection Regulation (DSGVO) simplify laws and impose penalties for infringements.

With the new regulations, it is becoming increasingly important to ensure that software and networks are free of security vulnerabilities. The study of ethical piracy can benefit various functions in this area. Network advocates and software developers can learn to identify and protect against common vulnerabilities. Managers and strategic planners can benefit from exploring common attack methods and their implications by incorporating this information into risk management plans.

## *Professional development*

The gap between the need for qualified cybersecurity workers and the pool of available talent is significant and growing. In the United States, there is an estimated 350,000 cybersecurity jobs, and this number is expected to increase tenfold by 2021. While this is generally bad news, and for companies seeking employees and employees, it is positioning itself in this field to retain its cyber security talent.

Learning ethical product piracy is a great way to get into cybersecurity and position yourself to take advantage of skills shortages (and the associated pay gap). Whether you are looking for a first job, specializing and progressing in a related field, or thinking about changing jobs, learning ethical hacking can be a good place to start to get the job you want.

## *Find Work*

If you're looking for a beginner position in a new field, you generally compete with other entry-level candidates with little or no practical experience. Anything you can do to prove your skills can differentiate you from other candidates and help you get the job done.

If you want to get into cybersecurity, it's best to get certified to prove your knowledge and experience. The various specializations and experience levels in cybersecurity are well represented by certifications.

One of the more common cyber security certifications is the Certified Ethical Hackers certification offered by the EC Council. The Certified Ethical Hacker exam tests the candidate's knowledge of the tools and techniques used by hackers, intruders and network lawyers.

Obtaining this certification demonstrates an applicant's willingness to work hard, and his success shows that he has the knowledge to complete the tasks required for his role. For those with limited experience in the field or who want to improve their skills, taking part in a bootcamp-style training can be a quick way to upgrade.

## ***Advantages of Ethical Hacking***

Most of the advantages of ethical hacking are obvious, but many are neglected. The benefits range from simply stopping malicious hacking to hindering national security breaches. The benefits include:

Combatting against terrorism and national security attacks

Having a computer system that blocks malicious hackers from gaining access

Having satisfactory preventative methods in place to check security breaches

## ***Disadvantages of Ethical Hacking***

As with all types of ventures which have a darker side, there will be unscrupulous people presenting hindrances. The possible drawbacks of ethical hacking include:

The ethical hacker using the information they gain to do spiteful hacking activities

Allowing the company's commercial and banking details to be viewed

The likelihood that the ethical hacker will create and/or place malicious code, viruses, malware and other destructive and hurtful items on a computer system

Massive security breach

These are not popular; however, they are something all companies should contemplate when using the services of an ethical hacker.

# CONCLUSION

Many people would be confused to read about the benefits of ethical hacking. To them, such a concept does not exist as hacking in itself is automatically viewed as unethical or illegal. Indeed, hacking is normally all about the breaching of barriers that have been put in place for the protection and security of the people. So to talk in terms of benefits of such acts is naturally quite alien to people (at least initially.)

Initially hacking really was all about the breaking of laws and accessing information that should not normally be accessed by certain groups of people. But life is never as black and white as we may first perceive. As such, it will come as a surprise to a good number of people that several major computer companies such as IBM, Microsoft, and Apple all have a large and dedicated team of hackers. Yes, you read that right.

They are not, however, breaking any laws so far as anybody can tell. No, these types of hackers are there for entirely good reasons. They are used as security testers for all sorts of programs. Basically, whenever a company comes up with a program, they'll usually bring it to their team of hackers who will then have a go at it ("hack") to see how many holes in security the program has.

They will see if the program can be exploited in anyway and then return it to the programmers along with a list of the vulnerabilities found. This is just one of the benefits of ethical hacking. The program can then be fixed, or strengthened, and sent back again to the hackers to confirm whether there are still any problems with it.

The aforementioned is just one example of the benefits in carrying out hacking. Did you know that there are actually courses being taught on this subject as the demand for hackers has actually increased? As the world becomes ever more reliant on computers, the potential damage that can be caused by a hacker, or groups of hackers, has grown to whole new levels. This is not something large companies can afford to ignore.

As such, learning how to be a hacker can lead to a very promising career indeed, working for one of the many major companies. As discussed, there are several good reasons for ethical hacking to be carried out "in-house" and all of them can help companies potentially save millions of dollars, and minimize the risk of ruining their hard earned reputation with their customers and peers. It is not only the companies who benefit but the people who buy their programs as well.

A team of good hackers can make sure that a program is as safe as possible, making the work of any would-be hacker that many times harder, often forcing them to move on to easier targets. This makes sure that any programs in wide circulation will rarely be tampered with and help protect the privacy and integrity of the computers of people all around the world.

In business, infrastructure equals money. In order to scale, you need a flexible infrastructure to handle the growth. With that said, when centralized infrastructure turns into bureaucracy and slow response, the company becomes lethargic. Hacking work examines these problems from the workers standpoint and outlines things you can do to get your work done by working smart.

Why is this important to me?

I am not doing this summary to waste your time. It is my vision to provide concise action steps that you can adopt right now to reach your entrepreneurial goals. Most companies today trust their vendors and customers more than their employees. This is a real problem because brilliant results require team work and you cannot have a cohesive team if there is no trust. Companies want transparency and centralization similar to command and control systems. This is not a bad thing until it takes a sales man 2 hours to enter an order or if the company blocks Facebook, twitter and LinkedIn. Stupid actions like this kill results.

Results are the name of the game. If you do not get results NOW, you are dead. The hub and spoke model for business is not a bad model just as long as the spokes have autonomy to deliver to the customers and are not tied up by bureaucracy.

Hacking work is broken down into four sections. For the sake of time, I will highlight one point from each section.

1. Engaged Team Members - This one point sums up the whole book and separates great businesses from crappy ones. Engaged team members are four times more productive and profitable than disengaged team members. This statistic if focused on can transform any business.

2. Slaves to Infrastructure - I understand the need for procedures and infrastructure because you cannot scale without it. With that said, I know that larger companies handcuff their employees with ridiculous rules and procedures that ultimately kill the creative spirit. Hacking Work is all about working around these ridiculous rules and procedures. A simple example of this would be locking down file transfer access from one

computer to the next. People today can have access to everything outside their work from their phone. Having stupid policies in place to limit creative freedom for the illusion of security is bad policy.

3. Three Types of Hackers - Black Hacks are the ones that steal, cheat and create havoc. These are the people who have given hacking a bad name. This book does not advocate black hacks. Grey Hacks and White Hacks are what are necessary to get the job done in a more efficient manner. These types of hacks are simply clever work around that save an enormous amount of time and allow workers to use their creative freedom for profit and customers loyalty.

4. Clarity - This one is a big deal. Take a look at the stats: one, three of the top five time wasters all relate to communication. Two, information in companies doubles every 550 days. Three, once every three minutes, the average cube dweller accepts an interruption and shifts her focus, consuming 28% of the day. Creating clarity and simple communication and information sharing networks can cure all of this.

**Do not go yet; One last thing to do**

If you enjoyed this book or found it useful I'd be very grateful if you'd post a short review on it. Your support really does make a difference and I read all the reviews personally so I can get your feedback and make this book even better.

Thanks again for your support!